D0541090

Lifeboat
Heroes

Lifeboat
Heroes

OUTSTANDING RNLI RESCUES
FROM THREE CENTURIES

Edward Wake-Walker

Haynes Publishing

*In memory of those who have given their lives
to save others at sea.*

First published in March 2009 by Haynes Publishing.
In association with the Royal National Lifeboat Institution.

RNLI name and logo are trademarks of RNLI used by
Haynes Publishing under licence from RNLI (Enterprises) Ltd,
Registered Charity No. 209603.

*The photographs used in this book are all from the RNLI archive.
Credit has been given where it is known, but the RNLI would be
grateful for information if any errors or gaps remain.*

A catalogue record for this book is available
from the British Library

ISBN 978 1 84425 545 0

Library of Congress control no. 2008939602

Published by Haynes Publishing, Sparkford,
Yeovil, Somerset BA22 7JJ, UK
Tel: 01963 442030 Fax: 01963 440001
Int. tel: +44 1963 442030 Int. fax: +44 1963 440001
E-mail: sales@haynes.co.uk
Website: www.haynes.co.uk

Haynes North America Inc.
861 Lawrence Drive, Newbury Park,
California 91320, USA

Printed and bound in Great Britain

CONTENTS

LIST OF COLOUR PLATES

ACKNOWLEDGEMENTS

I am grateful to many people at the Royal National Lifeboat Institution, not least James Vaughan who suggested I set about the task of selecting and researching 16 of the most outstanding lifeboat rescues of all time. Others at headquarters have helped me in my research at different times, including Joanna Bellis, Eleanor Driscoll, Derek King and Nathan Williams. The honorary librarians Barry Cox and Peter Morman have given me great assistance, as have many volunteers at the lifeboat stations I have visited in my search for material. I am also particularly indebted to Chris Price who forfeited a precious weekend to check the proofs on behalf of the RNLI.

One is always grateful to other writers and chroniclers when compiling a book of this nature, but I would like to give special mention to Michael Sagar-Fenton whose definitive and moving account of the Penlee lifeboat disaster, *Penlee, the Loss of a Lifeboat*, provided invaluable detail and local knowledge.

Of course, there would be no stories to tell without the willingness of volunteer lifeboat communities throughout the UK and Republic of Ireland to respond so wholeheartedly to news of distress at sea. Although this book concentrates on some of the most momentous events in the RNLI's history, we should all recognise and be grateful for every lifeboat launch, whether the incident is major or minor and regardless of whether it ends in triumph or disappointment. It is enough to know that there is always someone ready and able to go when the need arises, even in a gale and at dead of night.

INTRODUCTION

The Royal National Lifeboat Institution (RNLI) has awarded the Bronze, Silver or Gold Medal for deeds of outstanding bravery on more than 2,000 occasions. How do you select just a handful of these to represent nearly two centuries of heroism on the high seas?

If this book is to live up to its title, it must give an account of how the legendary names in lifeboat history earned their status. Men such as Hillary of Douglas, Fish of Ramsgate, Blogg of Cromer, Cross of The Humber, Evans of Moelfre, Bevan of The Humber and Clark of Lerwick secured their place in these pages because the story told is an example of just one of a string of medals they have won in their lifetime. There are others, perhaps not so nationally famous, such as Henry Alexander Hamilton of Skerries, Andrew Noble of Fraserburgh and Rod James of Hayling Island, who nevertheless were recognised by the RNLI on more than one occasion and deserve a share of the limelight.

Other criteria also helped to shape this collection. To explain how the lifeboat service developed and survives so successfully to this day as a largely volunteer organisation, relying on people's donations rather than the government for its funding, it was important to choose rescues at regular intervals throughout the RNLI's history. Also, to show the universal influence of the charity, incidents from all corners of the UK and Republic of Ireland are included. It surprises many people that the RNLI has been able to impose such high operating standards on groups of local volunteers who, by their geographical remoteness, might not normally be the most natural followers of head office rule books. The role of the lifeboat inspector, outlined in the story of Captain Charles Gray Jones's action-packed year of 1874, was, and still is, crucial in maintaining those standards.

Penzance lifeboat to the rescue in May 1888. Four men were saved when the *Jeune Hortense* was stranded in Mount's Bay, Cornwall. *(RNLI)*

Then there were the incidents that, by their very enormity and tragedy, affected the fortunes of the Institution. With the wreck of the *Indian Chief* off the Kent coast in 1881, there were victims and there were heroes. The survivors' story, and that of the Ramsgate lifeboat crew who saved them, was graphically relayed in the national newspapers of the day and brought the work of lifeboatmen into the consciousness of people across the land. When the *Mexico* ran aground in the Ribble Estuary five years later, the heroes became the victims when two lifeboats capsized and 27 men were drowned. This time the Victorian press and public began to ask if enough was being done to safeguard lifeboat crews.

Procedures about the decision to launch a lifeboat were reviewed and lifeboats were designed with greater stability. The disaster also inspired the first street collections for the lifeboats, bringing the appeal to modern industrial cities, often far from the sea.

Nearly a century later, just before Christmas 1981, when the *Union Star* turned turtle and Penlee lifeboat was smashed to smithereens under a Cornish cliff, killing everyone involved, again the nation woke up to the brutality of the sea and the bravery of those prepared to face its fury. Sympathy for the families of lifeboat crews provoked an onrush of unsolicited donations to the RNLI, allowing an accelerated programme of boat building, which provided more powerful and responsive lifeboats. These gave their crews a much better chance of survival when working close to the shore or alongside a floundering vessel in the worst of weather.

Lifeboat Day in London in the 1930s. Widespread public support of the lifeboat service only began in the late 19th century after two lifeboats were lost in the Ribble Estuary in 1886. (*London News Agency Photos Ltd*)

Henry Greathead, designer of the 'Original' lifeboat in 1789. (*RNLI*)

The 'Original' lifeboat. Thirty-one of these self-righting vessels were built for service around the coast of Great Britain and Ireland in the late 18th and early 19th centuries. (*RNLI*)

It is a grim fact of life that only when the inadequacy of tools for a specific job becomes tragically apparent are people inspired to find a remedy. Most turning points in the history of the lifeboat service are examples of necessity spawning invention and, at times, intervention. When the ship *Adventure* came to grief in a storm at the mouth of the Tyne in 1789 and thousands watched as her crew drowned, one by one, a committee of South Shields citizens resolved to find a boat which could survive such conditions and offer a chance of rescue to such casualties. The resultant self-righting design by Henry Greathead, the 'Original' lifeboat, replicated and exported to other coastal communities, became the vehicle which would inspire the veteran Manx lifesaver, Sir William Hillary, to campaign for a National Institution for the Preservation of Life from Shipwreck, as the RNLI was christened at its foundation in 1824.

A further disaster at the mouth of the Tyne in 1849, this time to a lifeboat similar to the 'Original', when 20 of her crew were lost during a rescue attempt to a wrecked brig, roused the RNLI out of the torpor that had overcome it only a few years after its establishment. Under a new president, the Duke of Northumberland, there was determination to update lifeboat design and to produce boats that could be sailed as well as rowed and whose stability, buoyancy and self-righting ability would encourage their use throughout the land.

When the first motorised lifeboat was introduced in 1905, again on the Tyne, there was effectively a mutiny by the Tynemouth crew who would no more trust an internal combustion engine than put the devil in charge of their boat. Only nine years later the superiority of motor lifeboats was proved beyond any argument by men from the same station who steered their latest petrol-fired boat 45 miles down the coast to Whitby in a wartime blackout and a storm. The Whitby pulling lifeboats had had to abandon heroic attempts to rescue the last survivors from the wrecked hospital ship *Rohilla*, but the Tynemouth boat got to her and brought the people to shore.

Fraserburgh lifeboat station has had more than its share of tragedy, as well as triumph, as Andrew Noble's story helps to illustrate. Twice, the town's lifeboat was turned upside down, in 1953 and 1970, both times with the loss of all but one of the crew. It was these disasters to non-self-righting Watson class lifeboats, added to another in 1969 when the Longhope lifeboat capsized and lost her entire crew, which convinced the RNLI that every all-weather boat built from then on should be able to right herself. They have remained so to this day.

A modern Tamar class lifeboat rights herself in seconds. A series of capsizes between 1950 and 1970 convinced the RNLI that self-righting should be inherent in the design of every all-weather boat. (*RNLI*)

Inflatable lifeboats, introduced in the 1960s, gave the RNLI a quicker response time and offered a more versatile means of reaching recreational sailors. *(RNLI)*

The introduction of an inshore lifeboat fleet cannot be attributed to any single event, rather a growing perception in the late 1950s and early 1960s that the increasing number of yachtsmen and others seeking pleasure from the sea needed a faster response than could be offered by heavy, 9-knot wooden lifeboats. The story in this book of the rescue of the people aboard the sail training ketch *Donald Searle*, floundering in shallow water off Hayling Island, gives as good an example as any of the indispensable attributes of an Atlantic 21 lifeboat with its shallow draft, its forgiving inflated tubes, its speed and its manoeuvrability.

There is a common thread that runs through all these stories and hundreds of other equally meritorious RNLI rescues for which limited space precludes a mention. The thread is one of a team of volunteers pitting their strength and ingenuity against wind and waves for the chance of saving life. Modern lifeboat crews may no longer be made up of a few local fishing families and will, instead, be men and women drawn from every trade and profession practised within the vicinity. They will each have received specific and detailed training, no longer chosen for their skill and strength as an oarsman. Unlike some of their Victorian predecessors, their family's next meal will not depend on the small allowance they are given for the time they spend at sea. And on the whole, thanks to today's equipment and technology, they will usually spend that time at less risk and in less discomfort.

The sea, in its worst mood, is still capable of testing the skill, nerve and stamina of a modern lifeboat coxswain and his crew to just the same extent as it did the man at the helm of an open pulling lifeboat. The problem of putting your boat alongside another wallowing or stranded vessel is as frightening a challenge as it ever was. And the determination and strength required by a crew to dislodge limpet-like survivors from the railings of their doomed ship has been a constant requirement throughout the centuries.

The question that remains at the end of all these stories of extraordinary bravery is: why do they do it? No one has ever become rich out of volunteering for the lifeboat. In fact, nowadays, where time is money, crewmembers afford long hours to train and exercise as well as to be available when the call comes. Whatever the motivation, be it a sense of duty, of tradition, of belonging to a team where winning is saving someone's life, there is still no shortage of men and women keen to join their local lifeboat station. Perhaps this book will help some of the newest recruits to understand what it will mean to serve the RNLI when sea and weather throw down their severest challenges.

1

DOUGLAS, ISLE OF MAN, 10 DECEMBER 1827

Sir William Hillary, founder of the RNLI, earns his first Gold Medal for bravery when he and his son Augustus take the Douglas lifeboat out to the Swedish ship *Fortroendet*, aground on St Mary's Isle, and rescue all 17 crew and passengers.

The Isle of Man had its Governor General, the 4th Duke of Atholl, to thank for becoming one of the first places in the British Isles to operate a lifeboat. The duke, an active member of the Royal Humane Society, ordered the lifeboat to be kept in readiness for the all too frequent shipwrecks that plagued the east coast of the island at the turn of the 19th century. The lifeboat, one of the 31 built to the same plans as Henry Greathead's 'Original', arrived in Douglas in 1802, although, along with most of the other original 31 lifeboats, little was recorded about her use. It is known that the boat was eventually lost in December 1814 after she broke from her moorings in a gale and was dashed to pieces at Douglas Head.

By then, however, the tradition of giving help from the shore to endangered seafarers had become well established and it was against this background that one of the island's more colourful inhabitants would eventually earn his place in history. Sir William Hillary, Bt, had settled in Douglas after losing his own and his wife's fortune in the patriotic, if rashly extravagant, formation of a private army to supplement the war effort against Napoleon. The semi-autonomous status of the Isle of Man, negotiated by the 3rd Duke of Atholl when he sold his family's sovereignty of the island back to the crown in 1765, provided Hillary with the legal refuge he sought from the debts and the wife he had abandoned in England.

Hillary, who was brought up as a Quaker, had a natural concern for the poor and dispossessed and a soldier's instinct to take personal action when that seemed to be required. When, in December 1822, three fishermen from Castletown on the south of the island drowned while rescuing the crew of a Royal Navy brig, the *Racehorse*, Hillary successfully campaigned for the Admiralty to provide their widows and children with a serviceman's pension. Earlier that same year he had persuaded some retired naval officers and local fishermen to join him in launching a number of frail rowing boats into Douglas Bay in a storm and saved six different ships from destruction.

Sir William Hillary, founder of the RNLI and three times Gold Medal winner for bravery. (*RNLI*)

Hillary had offered all the fishermen who had accompanied him that day a cash reward from his own fast-dwindling funds, knowing that they could not afford to risk leaving their families penniless, should the rescue mission end in disaster. By now Sir William's grand vision for the future of lifesaving had begun to take shape. The establishment of a national institution, funded perhaps by the Admiralty, which rewarded local rescuers for putting out to those in peril from shipwreck and which provided boats designed specially for the purpose, became his goal.

Again with his own money, he printed 700 copies of a pamphlet in 1823 entitled, *An Appeal to the British Nation on the Humanity and Policy of Forming a National Institution for the Preservation of Lives and Property from Shipwreck*. These he circulated to the many influential and aristocratic friends he had made in his days as equerry to Prince Augustus Frederick, the ninth child of George III.

Although considered a very worthy aspiration, no state-funded body saw it as its responsibility to adopt such an undertaking. Eventually Hillary's cause was taken up by Thomas Wilson MP, who reissued the appeal to wealthy philanthropists and whose influence and patience brought together the historic meeting in London of churchmen, politicians and financiers, that formalised the foundation of the National Institution for the Preservation of Life from Shipwreck as a charity on 4 March 1824.

Hillary was rightly regarded as the founder of the Royal National Lifeboat Institution, as it later became known, but his influence over it was always going to be limited by his enforced exile in the Isle of Man. Instead, equipped with the first lifeboat to be ordered by the Institution, he threw himself wholeheartedly into the business of running an exemplary outpost of the organisation at Douglas.

Sir William had two children, a twin son and daughter, who, after his escape across the Irish Sea in 1808, would become strangers to him. Scars from the broken marriage had healed sufficiently that in the summer of 1827 his son Augustus, a 27-year-old captain in the 6th Dragoon Guards, accepted an invitation from his father to visit the Isle of Man so that the two might get to know each other. It turned out to be a longer visit than originally planned, partly because father and son got on extremely well, but mainly because Augustus fell in love with the daughter of one of the island's judges, Deemster John Christian. During his courtship of Susanna Christian, Augustus Hillary found himself in a position to put his credentials as a worthy future husband and son-in-law beyond any doubt.

It was on 10 December 1827 that the island awoke to the sound of a brutish gale and heavy rain beating against the windows. As ever, when the wind blew hard from the east the people of Douglas kept an anxious eye out to sea for any vessel struggling against nature to keep clear of their rock-strewn lee shore. There was just such a ship, although they would not see her yet. She was Swedish, the *Fortroendet* of Karlskrona, carrying cargo and three passengers from Marseilles to Glasgow.

Her master, Andrew Kerman, realising the weather was deteriorating as the southern tip of the Isle of Man hove into view, knew that he needed to find shelter. His aim was to make for Douglas, but he first made a short stop at the small south-eastern port of Derbyhaven, to take on four extra crewmen and a local pilot as this was a dangerous and unfamiliar shore.

As the ship fought her way round Douglas Head in the strengthening gale, it became painfully obvious to the captain and pilot that the bay would offer them precious little shelter with the direction that the wind was now coming from. But to try to reach the harbour entrance or even to continue to sail northwards was too risky, so the only choice was to weigh anchor in the bay and hope that the gale would abate.

Any relief the skipper might have felt when the anchor held vanished moments later as a huge sea forced his ship's bow upwards, snatching the cable taut. All aboard felt a dull shock and immediately the vessel swung round, broadside to the waves, and began drifting unfettered towards the shore. Now there was nothing for it but to hoist enough canvas to make an attempt on the harbour entrance. This meant an approach with wind and seas full on the beam; the *Fortroendet* heeling over at an alarming angle, her crew clinging to the deck for dear life. Captain Kerman held this course for a few minutes until, fearing imminent capsize, he steered head to wind and ordered his crew to let go the spare anchor.

By now a crowd had gathered on the foreshore at Douglas, each rain and spray-soaked onlooker exchanging theories with their neighbour on what the captain should do next to save his ship or what his likely fate would be. Sir William Hillary and his son were among them but they, rather than watch a shipwreck unfold, were working out ways that they might help the captain. If a

A 20th-century aerial view of Douglas Harbour, Isle of Man. The bottom left of the picture shows a rocky island upon which a refuge stands, built by Sir William Hillary for survivors of ships wrecked on these rocks.

line could be put aboard the ship from the shore, she could perhaps be hauled into the harbour. Hillary ordered a mortar apparatus to be made ready for that purpose while he and his son made for the lifeboat. If the mortar failed to deliver a line, then maybe the lifeboat could take one out to the ship.

Just as Hillary, his son and some retired naval officers were clambering aboard the lifeboat, a cry went up from the onlookers; the Swedish ship had once more broken her anchor cable and was drifting through the breakers towards the off-

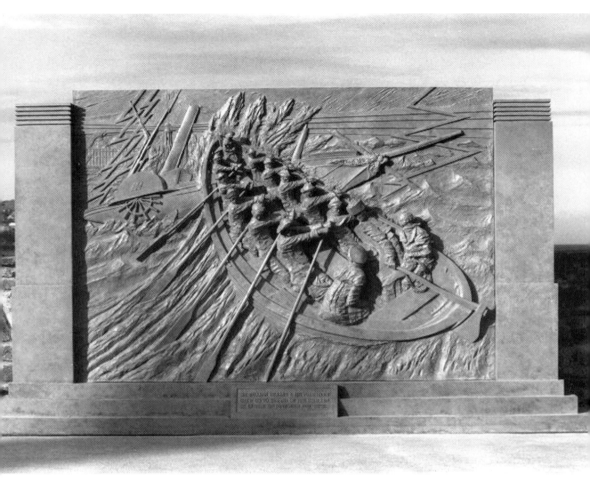

One of Sir William Hillary's Gold Medal rescues (to the *St George* in 1830) depicted in Michael Sandle's bronze sculpture on the seafront at Douglas. (*Michael Sandle*)

lying St Mary's Isle. They saw her progress suddenly halted as she grounded and then heeled over precariously on to her side. Her rudder had been torn away and water began to flood in through a gash in her hull.

Now all able hands ashore set about getting help to the crew and passengers. As well as the lifeboat, a ship's boat from the Royal Navy cutter HMS *Swallow* and two harbour boats, ill-suited for the conditions, were making their way out towards the wreck. The lifeboat, ahead of the rest, headed out to sea beyond the wreck, and when Sir William gauged that he was directly upwind of her, he let go an anchor and veered down upon her, his oarsmen struggling for control through the heavy breakers.

The lifeboat arrived intact under the lee of the ship and all 17 people aboard her dropped with their luggage into the lifeboat. The considerable extra weight now made the lifeboat far too low in the water for her to attempt a run in to shore. Fortunately, the Royal Navy's boat had got close to the wreck and so Hillary steered in her direction. Somehow the two boats avoided serious damage as they came together, their oarsmen struggling to keep head to sea while also avoiding hitting their opposite number in the other boat with the blade of an oar. A number of the survivors were bundled into the Navy boat and the two vessels headed for the shore, which they reached without further mishap.

Sir William wasted little time, once back ashore, to record his adventure at his writing desk and to send the report to the Committee of the Shipwreck Institution. Here was just the sort of successful lifeboat rescue that he was determined the organisation should be facilitating throughout the land. When the committee met to review their correspondence the following month, they were clearly impressed. Three years previously they had awarded him the Gold Medal as founder of the Institution, now they decided he should receive a second for his bravery. (Sir William Hillary would go on to win two more Gold Medals for bravery, both in 1830 when he was nearly 60.) Three Silver Medals were also awarded for this rescue: one to Lt William Strugnall who was aboard the Royal Navy boat, one to Robert Robinson who was aboard the lifeboat and one to Augustus Hillary, no doubt to the great delight of his father and his fiancée.

The view across to Lambay Island from Rush, the waters through which Henry Hamilton and his Skerries lifeboat crew battled to reach the *Tregiste* in November 1858. (*Edward Wake-Walker*)

2

SKERRIES, COUNTY DUBLIN, 17 NOVEMBER 1858

Henry Alexander Hamilton, honorary secretary of Skerries and Balbriggan lifeboats and the only Irishman to win a Gold and two Silver Medals for gallantry, saves the crew of the Austrian brig *Tregiste* in a rescue mission that lasts nearly two days.

On the wall of St George's Church in the coastal town of Balbriggan, which lies some 20 miles north of Dublin, there is a brass plaque which reads:

In loving memory of Henry Alexander Hamilton DL of Hampton in this parish, second son of Henry Hamilton of Tullylish, County Down, born 5 October 1820, died 30 March 1901. Honorary secretary to the General Synod of the Church of Ireland 1871–1894; Diocesan nominator for Dublin 1875–1892. For heroism in saving life at sea he was awarded 8 medals.

Quit you like men, be strong.

Research has so far not yet revealed the provenance of five of those eight medals, but three of them, a Gold and two Silver Medals were awarded by the RNLI.

Henry Alexander Hamilton was a member of an extended and highly influential Irish Protestant family whose ancestors were among the many Scots who crossed the Irish Sea in the early 17th century and settled in the north of Ireland. Henry's branch of the family settled in the Balbriggan area and built Hampton Hall, a large house and estate to the south of the town. They led and financed the development of the town with the establishment of a cotton mill in the late 18th century and by extending the harbour in the 1820s. One of Henry's cousins, George Hamilton, was first a Westminster MP and then became permanent Financial Secretary to the Treasury in Whitehall in 1859.

Henry, a bachelor all his life, pursued his interests closer to home, working not only as a pillar of the local church but as a magistrate and, from 1854 to 1900, as the honorary secretary of the RNLI lifeboat based at Skerries, about 2 miles south of Balbriggan. His obvious enthusiasm for the lifeboat service led to the opening of another lifeboat station at Balbriggan itself in 1874, where he was also the honorary secretary.

Shipwreck was a depressingly common occurrence on the Leinster coast in the mid-1800s. From 1850 to 1858 there were no fewer than 142 merchant shipping casualties on the stretch between Carlingford Lough and Dublin Bay, more than one a month on average. It was Henry Hamilton's personal involvement in one of these shipwrecks at Benhead, between Balbriggan and Drogheda, which is what led him to dedicate so much of the rest of his life to the lifesaving cause.

When the Whitehaven brig *Agnes* was driven ashore in a heavy gale early in the morning of 20 December 1853, there was no ready rescue service in the vicinity. Henry Hamilton was one of the group who had gathered on the shore to watch helplessly as the six-man crew signalled to them in desperation from the rigging. Hamilton and one or two of the others tried to launch local shoreboats to get out to the wreck but were hurled back on to the beach by the crashing surf.

A lifeboat was what they needed, but the nearest available one was in Dublin, more than 20 miles away. On many parts of the coast that would have meant the end for the crew of the *Agnes*, but here, as chance would have it, the Dublin to Drogheda railway ran right past the scene of the wreck. If the men could hold on long enough to their icy perch for someone to take the message into Dublin on the next train and for a lifeboat to be transported back on a railway carriage by return, maybe they could still be saved.

When the lifeboat finally arrived on the following day, Hamilton and five other men, including an American who happened to be on the spot, got the boat to sea but were beaten back to the shore by the strong winds and heavy seas. Their second attempt, made in darkness at 7.30pm was more successful and they got alongside the brig, by then, some 36 hours after she had run aground. Two

Henry Hamilton's legacy of a lifeboat station at Skerries was interrupted for some 50 years when this Liverpool class pulling lifeboat was withdrawn in 1930 after a motor lifeboat was stationed at nearby Howth. In 1981 Skerries reopened as an inshore station and today is equipped with an Atlantic 75 rigid inflatable lifeboat. (*RNLI*)

of the crew had earlier lost their fight to stay alive and dropped from the rigging, and a boy also died just minutes before the lifeboat reached him. However, the master and two seamen were still alive and were brought safely ashore.

Henry Hamilton's account of this incident, which was sent to the RNLI headquarters in London, brought about two things: a Silver Medal for gallantry for himself and a resolution to build a lifeboat shed at Skerries at a cost of

£98 18s 7d to house a 29ft James Peake-designed, ten-oared, pulling and sailing lifeboat at a further cost of £142 10s. The RNLI would pay half the cost and the local committee was to find the rest.

The lifeboat arrived in March 1854 and Henry Hamilton began his long service as honorary secretary. There was not a great amount of activity for the lifeboat station to begin with and it was, in fact, an incident in the summer of 1858 off Kingstown (now Dun Laoghaire), to the south of Dublin, which brought Henry Hamilton once more to the attention of the RNLI General Committee. They had been sent a letter from a Mrs Burden (who at the time wished to remain anonymous) and who was 'desirous to show her gratitude to the Almighty by presenting to the RNLI the sum of £300, to be employed by it in stationing an additional lifeboat on the Irish coast'.

Although there are tantalisingly few details, it seems that on 21 July 1858 the woman had been in a boat off the port of Kingstown when she fell overboard. Who should be on hand but the ever-alert Henry Alexander Hamilton who dived into the water and hauled her to safety. The committee minuted their grateful acceptance of 'the lady's munificent offer' and also resolved to award Hamilton with a second Service Clasp to his Silver Medal 'in acknowledgement of his gallant conduct'.

Little did he or the committee know that, barely four months later, yet another report of outstanding service, this time by the Skerries lifeboat, would be landing on the desk of the chief inspector of lifeboats in London. It was on 14 November 1858 that an Austrian brig, the *Tregiste*, carrying a cargo of coal from Troon to her home port of Trieste, was caught in a violent easterly gale and ran for shelter under Lambay Island, which lies 2 miles off the coast, north of Dublin.

By the next day, with the gale still raging, the ship had dragged her anchor half way across the sound between the island and the mainland. As the vicious rocks on the Portrane peninsula loomed ever closer, the English pilot on board the *Tregiste* strongly advised her skipper to cut her two masts away to give the ship less resistance to the wind. The captain agreed to this drastic and hazardous action but, unfortunately, he and one of his crew were severely injured as the masts fell.

Word of the brig's predicament soon reached Henry Hamilton and, probably cursing the fact that once again the emergency was far from the nearest lifeboat, he began to organise horses and Coast Guard men to transport the lifeboat on

The harbour at Skerries. When the alarm was raised about the plight of the *Tregiste*, the lifeboat had to be transported on her carriage 5 miles south along the coast to Rush. (*Hunting Aerofilms*)

her launching carriage to Rush, a fishing village some 5 miles to the south which would be in sight of the ship in trouble.

At 2pm on 15 November the lifeboat launched from the strand to the south of Rush. Chief Boatman of Coast Guard, Joseph Clarke, took up his position as coxswain and beside him, ignoring any convention that the honorary secretary remain ashore, was Henry Hamilton. For the men at the oars it was an impossible task to get across the bay to reach the *Tregiste*. They were rowing broadside on to the weather and as they got into deep water, seas began to fill the boat. Hamilton later recounted: 'Twice, many of us were nearly washed out of the boat, the green water falling at times unbroken on the top of us.'

After two hours the lifeboat was still a long way from her objective; the crew were utterly exhausted and Hamilton realised that if she didn't now turn for the shore, she could well end up on the rocks herself. They headed eastward, downwind, for the entrance to Rogerstown Stream. Crossing the bar, they narrowly avoided broaching and eventually arrived in calm water at 5.30pm.

Fortunately there was a Coast Guard watch-house in the vicinity and so the lifeboat was moored in the river and the crew trudged wearily towards it for shelter. Hamilton's plan was to obtain food and dry clothing for his men while they all remained at the watch-house, keeping a close eye on the *Tregiste* in case her anchor dragged or the cable parted. With the storm as strong as ever, no muscle-propelled vessel had a chance of reaching her in her present position; the only hope was to wait for a lull in the weather.

Miraculously, when dawn broke with the wind still howling off the sea, the hull had not moved any closer to the rocks overnight. The men camping in the Coast Guard lookout watched with some hope as a little later that morning a large steamer hove into view in the bay. She had been summoned from Holyhead to try to take the men off the ship and had survived a torrid crossing of the Irish Sea. But the seas which continually broke over her thwarted her captain's efforts to veer down on the casualty to get alongside and eventually he had to slip his cables, leaving his anchors on the seabed, and make for the safety of Kingstown Harbour. As the second night set in, Hamilton and his men could only watch as the *Tregiste*'s crew worked feverishly at the pumps while waves broke over them, if anything, more heavily than before.

Then, at about 3 o'clock the following morning, the men in the watch-house sensed a change in the weather. There were still angry gusts pummelling the roof but they were less prolonged and lacked their earlier menace. Hamilton decided it was time to try the lifeboat again, and at 4.30am on the 17th he and his crew were heading out into the bay once more. For the first hour it seemed that they should never have set out as seas met the lifeboat head-on with as much violence as before. Gradually, though, they seemed to come through the worst and after two-and-a-half hours of agonising effort, the oarsmen reached the *Tregiste*. They now had to find more strength to haul the survivors over the stern of the ship and into the lifeboat, timing the moment to perfection as the decks of the two vessels drew level.

All 13 people aboard, including the injured captain and crewmember, were safely transferred and the lifeboat was able to land them all at Rogerstown at 9.30am.

The ship survived the storm and was eventually towed to Kingstown for repair. Neither the rescuers nor her crew could ever have been certain of such an outcome and it certainly did not detract from the admiration afforded Hamilton and his men on their safe return. The RNLI presented Henry Hamilton with their highest honour this time, the Gold Medal for gallantry, placing him among the elite in the history of the organisation to have won so many accolades.

By the time Henry Hamilton retired as the honorary secretary of Skerries at the age of 80, the station, which he had been largely responsible for establishing, had launched its lifeboat on 18 rescue missions and saved 52 lives in the process.

The 19th-century lifeboathouse and slipway at Appledore. (*Grahame Farr Archives*)

3

APPLEDORE, 28 DECEMBER 1868

Appledore lifeboat is launched twice into the raging surf in a north-westerly gale to rescue the crew of the grounded Austrian barque *Pace*. On the first attempt, nine people are brought ashore at the expense of the lifeboat's rudder, which is torn off as the two vessels are thrown together. Joseph Cox, the Appledore coxswain, in spite of an injury received in the same collision, puts out to sea again to try to reach the rest of the ship's crew. This time the lifeboat capsizes just short of the wreck, but all her crew manage to get back on board and return to shore.

When the wind is blowing hard from the north-west into Bideford Bay, it is a very inhospitable place for a ship to be under sail. Joseph Cox knows this only too well as he peers out across Bideford Bar where the rivers Torridge and Taw join forces to confront the unforgiving Atlantic. The view is one of angry white water stretching into the distance as currents and wind provoke the sea into a foaming frenzy over the sandbanks of the bay.

Cox has been coxswain of the lifeboat at Appledore for more than 16 years and has seen every kind of mishap in that time. Twenty-seven people from five different shipwrecks owe him and his crew their lives after rescues in weather not as threatening as today. Five years ago the RNLI saw fit to award Joseph Cox its Silver Medal 'for his long and gallant services', but he is hoping that they will not be called upon this raw December lunchtime as he heads home for a bite to eat.

He is scarcely indoors when he hears hefty knocking at his door. A breathless messenger from the Coastguard brings the news he hoped he would not hear:

two large sailing ships have been swept into Bideford Bay and are fighting what looks like a losing battle to sail clear of the shore. One, if not two shipwrecks are imminent.

The coxswain immediately gives the alarm, grabs an extra jersey, and sets off across the low sandy scrubland of Northam Burrows to where the lifeboat is kept, ready for launching either into the estuary or the open sea to the west. The lifeboat's name is *Hope*, not through sentiment, however appropriate, but in honour of her donor, Mrs Ellen Hope, whose clergyman husband had made the gift of a lifeboat his dying wish.

The boat is a 34ft self-righter which is light enough to be propelled by 6 oarsmen, but wide enough for 12 short oars when conditions dictate. She has been at Appledore for seven years and has proved herself an excellent boat, especially after two recent incidents. One was only last year when she reached the three-man crew of a sunken brig in heavy seas and they were plucked from the rigging and got to safety. The other was a year earlier when another brig came to grief inside the bar; the lifeboat was able to tow the ten survivors in their ship's boat through a mass of broken water to the shore.

Walking briskly along the rough road that leads to the lifeboathouse, Joseph Cox hears the sound of other men running up behind him and the more distant pounding of horses' hooves. The first to catch him up and clap a hand on his shoulder is his son, also Joseph and a regular member of the crew. Father imparts the scant information he has to his son by which time they and a dozen other men have reached the station, are donning oilskins and cork lifejackets and are preparing the lifeboat carriage for a launch.

The launchers arrive and set to work harnessing the horses to the carriage. In a matter of moments the lifeboat is out of the shed and trundling across the sands to the water's edge. Everyone on the beach can now see the two barques in difficulties. One of them, an Austrian vessel, the *Pace* of Fiume (nowadays Rizeka in Croatia), on her way home from Glasgow carrying pig-iron and coal, is desperately close to the bar, trying to claw her way out into open sea but losing ground with every breaking sea.

Horses, lifeboat and men move slowly along the beach, following the barque's southward drift until, at about 2pm, she comes to an abrupt halt and seas begin to break in white plumes over her. She is aground. The lifeboat crew scramble up into the boat, the carriage is hauled over a ridge of shingle and into the sea so

A late 19th-century launch of Appledore's No. 2 lifeboat across the beach at Northam Burrows. (*RNLI*)

that launchers and horses are half submerged in the roaring, icy surf. The oars are swung outboard at the ready, the coxswain bellows, the carriage tips and the lifeboat's bow plunges into the breakers.

Immediately the 12 oars begin to work as if driven by a single machine. The lifeboat crashes through one breaker, disappears completely in the trough beyond, then rears up to the angle of a ladder against a wall as she hits the following wave head-on. And all the time the oars keep up their perfect rhythm, each man knowing that if he were to miss a single beat, it could be enough to throw the lifeboat broadside on to the breakers causing an instant capsize.

Oars at the ready for the moment the carriage tips to launch the lifeboat. (*Grahame Farr Archives*)

No one on board hears the cheers from the shore as the lifeboat finally makes it through the surf into less broken water. Somehow they now have to get alongside the wreck, her hull only sometimes visible as seas sweep clean over her. The oarsmen summon up their dwindling breath to pull just close enough to the *Pace* for the lifeboat bowman to throw a grapnel, which, to everyone's relief, catches in the rigging. He heaves on the line to bring the lifeboat in close.

Joseph Cox, Snr, is concerned. The rigging is where he expected to see the survivors, assuming there are some; it is the only relatively safe place to be in these circumstances. Held fast to the wreck by the bow, the lifeboat is rising and falling with the waves and all on board are constantly deluged as the seas which pound the barque then cascade into the open lifeboat. If the *Pace's* crew are still on deck, wonders the coxswain, why has no one stepped forward to take the lifeboat's stern rope to bring her closer alongside?

He shouts; his crew shout, but there is no movement, no sound in response. Are they frozen? Unconscious? All dead? Surely that's impossible when they were under sail only a short while ago. Then they hear an angry shout coming from the for'ard end and moments later the pallid face of a boy appears at the ship's rail. Scarcely before anyone can move aboard the lifeboat, the boy has clambered on to the rail and has jumped. He lands in the bow of the lifeboat, the bowman just able to break his fall.

Try as they might, the Appledore lifeboatmen can make no sense of what the agitated boy is trying to tell them in his very foreign tongue. Five baffled minutes pass and then there is the sound of a stampede on the wooden deck above them. Eight men appear and, as if chased by a wild animal, jump over the side, miss the lifeboat and plunge into the sea. Swiftly and expertly the oarsmen manoeuvre so that others can lean over and haul each spluttering survivor into the lifeboat.

Just as the last man is coming aboard, the lifeboat is momentarily caught broadside to the seas and is driven hard against the barque's overhanging stern. There is a sickening wrenching sound and as the two vessels move apart, the lifeboat's rudder is floating flat on the surface of the sea, torn clean away from the stern. Of even more concern is the sight of the coxswain, slumped down in his position at the stern, his hands clenched tight together, his head bent forward as he rocks to and fro in obvious agony. As the two vessels came together, his torso was caught between the end-box of the lifeboat and the *Pace*'s side and squeezed with brutal force. Gradually he recovers his breath as the others look on anxiously. Then he unbends, coughs, takes three deep but uneasy breaths, looks at his son, nods, and pulls himself slowly to his feet to resume command. As he stands once more in the stern, his crew are astonished to see that sections of cork are hanging broken from his side and realise that his lifejacket must have saved him from being instantly crushed to death.

Each lifeboatman knows that eight men and a boy is too small a crew for a barque the size of the *Pace*, but they can get no answer from anyone else on board; neither can they get any sense from those they have rescued about how many of their shipmates are left aboard. Again they try to make their voices heard above the gale and breaking seas, urging them to join the lifeboat, but no one stirs.

Joseph Cox now decides that they have hung on long enough and it is time to get the survivors ashore. He shouts to the bowman to cut the rope holding them alongside and, with an oar thrust out over the stern, steers the lifeboat towards the beach. The boat careers forward unceremoniously in the surf, the oarsmen struggle in the pause before the next breaker to keep her stern to the seas, then she is picked up again and surges shorewards. Launchers dash into the shallows to take the bowline and the lifeboat is hauled up the beach. The coxswain tells the launchers to get the lifeboat back on the carriage for another launch as soon as possible; her job is only half done.

He has finally ascertained what has been going on aboard the grounded ship. Effectively, a mutiny has taken place, led by the boy now safely ashore. The Austrian skipper had assembled all his men under the shelter of the cuddy and ordered them to stay put, forbidding them to acknowledge the lifeboat's presence in any way. He believed that there was hope yet for his ship as the tide was beginning to fall. If the weather improved, there was a chance, he felt, she could be refloated on the next tide. Clearly the eight men who disobeyed orders and followed the boy into the lifeboat did not share his optimism. And the coxswain most certainly does not; he still fears greatly for the five men left on board.

There is relief for the exhausted oarsmen as the survivors are ushered away; they will not be needed again as there are at least enough men for a fresh crew foregathered on the beach. But for the aching Joseph Cox and his son and one other, John Kelly, there is more punishment to come. Their skill and experience are indispensable for the second mission to the *Pace*.

The replacement oarsmen prove themselves to be as strong and expert as the first and, in spite of the damage to the lifeboat, she is pulled clear of the surf for a second time. This time Joseph Cox, Jnr, takes a turn to steer with the oar. In the fading winter afternoon light, he is visible from the beach, standing high in the stern, the makeshift tiller wedged under his right arm as the lifeboat makes her bucking progress out towards the wreck. As the lifeboat reappears after a huge wave sweeps her end to end, the figure of the coxswain's son is no longer there.

He is in the water. The wave has catapulted him over the stern end-box, his oarsmen watching helplessly as he disappears. Now, without a helmsman, the sea has finally won the battle for control and the *Hope* slews broadside to the waves. Immediately, she is bowled upside-down and every man on board is tipped head-first into the sea.

First to grab the boat after she has sprung back upright is young Joseph Cox. With the strength that comes with desperation, he heaves himself over the side and back on board. There are men all around him in the water; he grabs the wrist of the nearest to him and brings him half inboard. Now he can reach down his back and grasp the bottom of his lifejacket and bundle him into the boat. After a few minutes, 14 men have regained the lifeboat, all numb with cold and many still gasping for air, having swallowed quantities of seawater. Twice Joseph Cox counts the oilskin-clad figures as they move about the boat, searching for oars and attempting to restore order.

Fourteen men are aboard but they had put out with 15. Who's missing? Where's the old man? Where's the coxswain? He's not there, he must still be in the water! The men scour the seething surface of the sea and someone spots a rounded shape about 50 yards away. Only three oars are left in the boat after the capsize. With a huge effort and great skill, the lifeboat is coaxed towards the bobbing shape which is, indeed, the coxswain. A crewman reaches out at full stretch with one of the oars and, to the relief of all aboard, Joseph Cox, Snr, shows there is still life left in him by grasping the blade firmly and allowing himself to be pulled to the side of the boat and then aboard.

He is utterly exhausted and has sustained further injuries in the capsize. His son knows that he must be got ashore and that, with only three oars, any further rescue attempt would be impossible. As the oarsmen bring the lifeboat's bow round to head for the beach, he looks back at the *Pace* where five men, who have at last taken refuge in the rigging of the mizzen mast, watch dejectedly as the *Hope* pulls away from them.

Skilful oarsmanship and some good fortune bring the lifeboat safely back through the surf for a second time. Many more people still have gathered on the beach and there are more than enough volunteers ready to crew the lifeboat for a third launch. Among these volunteers are the crew of the Braunton lifeboat, stationed on the north side of the estuary. They had earlier tried to launch their own boat to the wreck but had been driven back. Now they have walked several miles and crossed the river to see how they might help the Appledore crew.

William Yeo, a banker from Bideford and a prominent member of the lifeboat committee, has taken command of the situation on the beach. Seeing the lifeboat return through the surf with only three oars, he has sent horsemen back to Appledore for replacements. They are now galloping back along the beach, each

Coxswain Joseph Cox, his hand on the helm, poses with other Appledore lifeboatmen aboard the lifeboat *Hope*, from which he nearly lost his life saving the crew of the Austrian barque *Pace* in December 1868. (*RNLI*)

carrying an oar, like lancers into battle, and they hand them, breathless, to the waiting volunteers.

But Yeo is not happy to let the lifeboat go out again. Already there has been a near disaster in a lifeboat that is far from intact. What is more, the tide has fallen a good 2ft, so the danger to the *Pace* is already less. She can be seen full of water, lying in a hollow in the sand, almost motionless. Soon men would be able to reach her without a boat and lead the survivors to safety.

So the lifeboat stays put and after a while a party of men wade out through the surf to the wreck. The captain and two of his men, numb with cold, clamber stiffly down from the rigging. With their arms around the shoulders of their rescuers, they stumble through the shallows to dry land. The other two men, who had stayed with their captain, have earlier lost their battle with the elements, dropped from the rigging and drowned.

Joseph Cox, Snr, as he is having his injuries inspected at the lifeboathouse, remembers with a start that he had originally been told of two ships in difficulties. What had become of the second one? He would not learn the full story until later that day back in Appledore. She was the *Leopard* of London, on her way to Gloucester from Sombrero Island in the West Indies. She eventually drove ashore at Westward Ho!, about 2 miles to the south-west of where the *Pace* ran aground. The rocket brigade could not get their line aboard in the fierce onshore wind, so David Johns, a Coastguard boatman and one of Cox's crew on his first launch to the *Pace*, volunteered to swim out with a line.

He made it out to the stranded ship but, in his attempt to board her, was struck on the head by a piece of wreckage and drowned. This tragedy did not deter another Appledore man, George Galsworthy, from trying the same thing again. He was more fortunate and clambered on to the ship with the line, enabling all her crew to reach safety.

Joseph Cox recovered from his injuries and remained coxswain at Appledore for another four years. His two launches to the *Pace* earned him a second and third Service Clasp to his Silver Medal. His son and John Kelly also received Silver Medals and all three were later awarded a silver cross of merit by no less a figure than the Emperor of Austria.

A low-water launch of the pulling lifeboat *Co-operator II*, which was stationed at Ilfracombe between 1893 and 1921.

4

NEWCASTLE, COUNTY DOWN, FEBRUARY 1874, AND ILFRACOMBE, DEVON, DECEMBER 1874

Captain Charles Gray Jones, RN, in his brief career with the RNLI as its Second Assistant-Inspector, finds himself winning the Silver Medal for bravery twice in a year, first aboard the Newcastle, County Down, lifeboat when four men are rescued from the rigging of a schooner wrecked in Dundrum Bay and then at Ilfracombe when he and the lifeboat crew save the crew of two brigs in the Bristol Channel.

The RNLI of the 1870s was a very different outfit from the one that, 20 years earlier, had been forced to go cap in hand to the government to save it from oblivion. Under the highly capable stewardship of its Secretary, Richard Lewis, the Institution's independent financial viability was restored and lifeboats were at last becoming the indispensable tool for communities that had been Sir William Hillary's vision when he founded the charity in 1824.

Between its foundation and 1850 (the year of Richard Lewis's appointment), the RNLI, or the 'Shipwreck Institution' as it was colloquially known, had degenerated to little more than a remote committee of titled gentlemen who would meet each month in London to confer financial rewards and sometimes gallantry medals on seafaring folk for their lifesaving deeds. It was very rare indeed for the rescues they recognised to have been carried out in one of the Institution's own boats, mainly because local men preferred to use their own locally designed boats and would shun official lifeboats, viewed by many as cumbersome or inappropriate.

At most lifeboat stations there was a boat, but not much else; no one to maintain her and often no one officially to take command at times when she might have been of use. The post-1850 General Committee, now containing naval officers with practical knowledge of seamanship, realised that to make the RNLI an active and effective rescue service it needed lifeboats designed to take local conditions into account. Moreover, there needed to be a retained coxswain at every station who was paid a fixed salary and who would be responsible for the boat's maintenance. They, in turn, would be supervised by full-time officers of the Institution, who made regular inspections of lifeboats to ensure that they were seaworthy, that they were always ready to be launched and that a crew was available at all times.

The official qualifications required to become an Inspector of Lifeboats in the early days of the reformed RNLI and the duties they needed to perform included the following:

QUALIFICATIONS

He should have followed the sea as a profession and must produce testimonials of his services. He should have paid special attention to boat work and be active and in good health. It is desirable that he should not be more than 40 years of age on his first appointment; and he will be called on to retire at 60 years of age, or earlier, should his activity become impaired.

He should be intelligent and able to speak at public meetings on subjects connected with the operations of the Lifeboat Institution. He should also possess a good address and, by patience, tact and temper, be able to win the confidence of the fishermen and the seafaring population and endeavour to obtain their zealous co-operation in the lifeboat service.

He should be able to write clearly and concisely and have knowledge of accounts and of keeping stores.

DUTIES

He will visit the coast in his district periodically and thoroughly inspect the lifeboat stations.

He will, as a rule, inform the honorary local secretary of his intended visits and times of arrival and, in conjunction with the latter, will examine and check all the documents requiring his notice; but this rule is not to interfere with his visiting the station at other times when he deems it desirable to do so.

He will take the boats out, when practicable, at least twice in each year.

He will move about amongst the crews of the lifeboats and confer with the coxswains as to the conduct and ability of their men, and should he find prejudice against the working of the system in any particular locality, he will endeavour to remove it.

He will correspond with and receive his instructions from the Chief Inspector; and all his reports are to be addressed to that officer who will pass them to the Secretary, together with a summary of their contents, for the information of the General Committee.

He will meet the Chief Inspector on his visits to the coast and call attention to any points which he may consider require the Chief Inspector's personal observation.

He will attend public meetings in his district called in furtherance of the objects of the Institution and encourage local interest in its favour to the best of his ability. He will also, in conjunction with the local committees and secretaries, use every effort to maintain and improve the interests of the Institution, financially and otherwise, in each locality.

He will always bear in mind that the Lifeboat Service in all its details is voluntary, and in his relations with the local secretaries and committees, and in his conduct towards the coxswains and crews, he should exercise the greatest conciliation, consistent with the necessary vigilance in maintaining the several lifeboat establishments in a state of thorough efficiency.

Contained within these brief paragraphs is the essence of the RNLI's extraordinary success as a national voluntary rescue service over the past 150 years. Any inspector joining the RNLI today would find remarkable similarities between these 19th-century instructions and his current job description. It has always been in the hands of a lifeboat inspector to balance the recognition that his workforce is voluntary with the discipline required in maintaining a seamanlike operation. That quintessential skill, arguably above all others, is what has made the RNLI so great for so long.

The hierarchy of the RNLI inspectorate in the 1870s allowed not only for a Chief Inspector and five District Inspectors, but also for an Assistant-Inspector and a Second Assistant-Inspector operating from the headquarters in London. Appointed to this latter post in February 1874, at an annual salary of £200, was one Captain Charles Gray Jones, RN. Although his employment with the RNLI lasted less than four years, he was to achieve, in his very first year with the Institution, a record which has never been repeated by a lifeboat inspector.

The job of Second Assistant-Inspector was clearly not a desk-bound one. Within days of his taking up his position on 5 February 1874, Gray Jones was dispatched on a tour of the east coast of Ireland visiting stations at Rogerstown, Drogheda, Dundalk, Newcastle, Tyrella and Ballywalter. The weather during his trip was clearly determined to give him a baptism by fire and when he awoke in his hotel bed in Newcastle, County Down, at daybreak on 26 February, he could hear the full force of a heavy south-easterly gale beating against his window.

The view of Dundrum Bay which greeted him was of a jagged grey and white mass of breakers heaving and elbowing their way towards the shore and crashing in shimmering plumes against the sea wall. As his eyes adjusted to the scene, he began to make out through the spray the shape of a ship's mast, a schooner with sails in ribbons and with precious little distance between her and the surf-strewn rocks to the south of the bay. The gale was blowing her straight at the shore and she would surely be a wreck within the hour.

The town of Newcastle had an exceptional record when it came to courageous deeds of lifesaving in Dundrum Bay. In the 50 years since the RNLI's foundation, the Institution had awarded 4 Gold and 12 Silver Medals to its citizens. Although a lifeboat had been stationed close to St John's Point at the north end of the bay as early as 1825, neither that one nor its eventual successor placed at Newcastle itself in 1854, had been used in any of these medal-winning rescues. Now, however, there was a chance for a lifeboat of the modern RNLI to prove herself and Captain Gray Jones was down at the lifeboat station like a shot.

James Hall, coxswain of the Newcastle lifeboat, would have found it difficult to tell his important visitor from head office that he would rather leave him on the shore for this mission, even if he had felt inclined to do so. The Second Assistant-Inspector was obviously determined to demonstrate his credentials as early in his RNLI career as possible and took command of the situation. The lifeboat, the *Reigate,* was hauled out of her shed and before long her crew was pulling hard on the oars to clear the hefty breakers, Gray Jones standing erect beside the coxswain in the stern.

Even before the lifeboat was afloat, the schooner, the *Rose* of Youghal, bound originally for Dublin from Bridgwater, was aground. The moment she struck, water poured in through her hull and over her deck and one of her crew was swept off his feet and carried overboard, never to be seen alive again. The four men who remained aboard were quick enough to make it to the fore-rigging as the ship disappeared beneath the surface.

When the lifeboat arrived, she made her approach on the leeward side of the wreck, negotiating a strong tide and a heavy cross-sea strewn with debris. In spite of the risks, she got alongside the rigging long enough for the men, who had been clinging on for their lives, to be taken aboard and returned without further mishap to dry land.

The story of this rescue was included in the report of his Irish sojourn, submitted by Captain Gray Jones to the General Committee of the RNLI at their March meeting. One can only imagine his reaction when they emerged from the meeting announcing that both he and Coxswain Hall had been awarded the Silver Medal for their exertions.

At the end of that same year, Charles Gray Jones was once again on his travels. This time it was the West Country in December and once again his report to the General Committee at the end of his trip gave them more to ponder than routine matters such as the condition of the boats, their equipment and the well-being of stations visited.

Ever the man to find himself in the right place at the right time, Gray Jones was in Bude on Cornwall's north coast on 6 December 1874 when a westerly gale sprang up. The smack *Charlotte* was caught out in it and driven helplessly into Widemouth Bay, about 2 miles to the south of the town. The vessel soon

Above and opposite: The launch of the pulling lifeboat *John Cleland*, which saw service at Newcastle, County Down, between 1917 and 1932. (*RNLI*)

became a wreck, but Gray Jones was among the people who had hurried down to the beach and, ignoring its icy temperature, waded into the crashing surf and helped to haul the smack's master out of the sea and to safety.

Ten days later he had progressed up the coast to Ilfracombe where stormy weather was, once again, threatening commercial traffic at sea. This time it was an easterly gale so that by the early morning of Wednesday 16 December, heavy seas had built up and were surging through the Bristol Channel, making the north-east-facing entrance to Ilfracombe harbour a frightening place to be. Already the Swansea steamer, *Henry Southan*, had abandoned her attempt to leave the harbour with a number of sheep and cattle aboard.

When the lifeboat was called out at Ilfracombe, the townspeople turned out to watch. (*RNLI*)

The *Broadwater*, the lifeboat in which Captain Charles Gray Jones earned his second Silver Medal in the same year. (*RNLI*)

Then news reached the town that an unidentified brig was brought up at anchor to the north-east, and was attempting to ride out the gale in an extremely exposed position. Before long she had disappeared from her mooring and lookouts in the town could see her running down the Channel, just off Ilfracombe, with her fore-topmast and all head-gear and head-sails dragging in the sea under her bows. Her maintop-gallant yard had apparently also gone and she was clearly out of control with the cliffs of Morte Point only about 3 miles ahead of her.

Once again, nothing could have stopped the Second Assistant-Inspector from donning a cork lifejacket and clambering aboard the town's lifeboat, *Broadwater*, as she was got ready for launching at the top of the slipway. It was a titanic struggle against a flood tide and the head-on gale for the men at the oars to clear the harbour. They eventually succeeded, where a steam vessel had earlier failed, and were able to hoist a sail as they turned eastward in pursuit of the runaway brig.

The brig, the Dublin-based *Annie Arby* with seven men aboard, was wallowing in heavy seas right under the cliffs to the north of Morte Point when the lifeboat caught up with her. Once they had manoeuvred their boat alongside, members of the lifeboat crew boarded the brig and set about cutting away the wreckage while others worked the remaining sails. Inch by inch the vessel wore clear of the

A 1936 gathering of Ilfracombe lifeboatmen. Some of the older crewmembers may well have begun their service aboard the *Broadwater*, which served the station until 1886. (*R.L. Knight*)

cliffs and, with the local expert knowledge of one of the lifeboat crew, she took a narrow passage through the rocks which litter Morte Point and was brought into the comparative shelter of Morte Bay where a safe anchorage was possible. The brig would eventually be taken in tow by a tug when the weather eased.

Unbeknown to the lifeboat crew, another brig, the *Utility* of Workington, had also run into trouble astern of them while they were chasing the *Annie Arby*. On their return to Ilfracombe they happened upon a ship's boat with five men aboard drifting helplessly off Rockham Bay. When they pulled the survivors aboard the lifeboat they were told that their vessel had filled and sunk as it struck the rocky foreshore and that they had barely had time to scramble into their boat.

The RNLI General Committee realised, when they considered his report, that they could not possibly ignore the fact that on his latest tour of inspection, the remarkable Captain Gray Jones had this time presided over the saving of a total of 13 lives, all in extreme circumstances. The award of another Silver Medal seemed the only appropriate course of action for them to take.

A plaque from the stern of the *Indian Chief*, which was wrecked on the Long Sand off the north Kent coast with the loss of all but 11 of her crew. (*RNLI*)

5

RAMSGATE, 6 JANUARY 1881

After two harrowing nights aboard the rapidly disintegrating barque *Indian Chief*, aground in a gale on the Long Sand off the north Kent coast, 11 men are rescued by the Ramsgate lifeboat. The lifeboat crew, led by Coxswain Charles Fish, have to survive a night on the water aboard the lifeboat in mountainous seas before they can find the wreck.

This grim description of the scene on Ramsgate pier on the afternoon of Thursday 6 January 1881 was given to readers of the *Daily Telegraph* following one of the greatest tests of endurance ever faced by survivors and rescuers alike in the history of the RNLI.

One by one the survivors came along the pier, the most dismal procession it was ever my lot to behold – eleven live but scarcely living men, most of them clad in oilskins and walking with bowed backs, drooping heads and nerveless arms. There was blood on the faces of some, circled with a white encrustation of salt, and the same salt filled the hollows of their eyes and streaked their hair with lines that looked like snow.

The first man, who was the chief mate, leaned heavily on the arm of the kindly-hearted harbour master, Captain Braine. The second man, whose collar-bone was broken, moved as one might suppose a galvanised corpse would. A third man's wan face wore a forced smile, which only seemed to light up the piteous underlying expression of the features. They were all saturated with brine; they were soaked with seawater to the very marrow of the bones. Shivering, and with a stupefied rolling of the eyes, their teeth clenched, their chilled fingers pressed into the palms of their hands, they passed out of sight.

As the last man came I held my breath; he was alive when taken from the wreck, but had died in the boat. Four men bore him on their shoulders, and a flag flung over the face mercifully concealed what was most shocking of the dreadful sight; but they had removed his boots and socks to chafe his feet before he died, and had slipped a pair of mittens over the toes which left the ankles naked. This was the body of Howard Primrose Fraser, the second mate of the lost ship and her drowned captain's brother.

The 11 men and one corpse were all that were left from a proud ship's company of 28 who set sail from Middlesbrough on New Year's Day aboard the three-masted barque *Indian Chief*, bound with a general cargo for Yokohama in Japan. An icy north-easterly gale had carried them down the east coast and the crew were no doubt looking forward to reaching more southerly latitudes on their long voyage to the east. By the early hours of Wednesday 5 January they were negotiating the tricky waters off the Thames Estuary where hidden sandbanks lurk, menacing any ship that wanders off her course.

Captain Fraser of the *Indian Chief* had just sighted the Knock Light when the wind shifted to the east and they were hit by a heavy squall of rain. Concerned that he was drifting towards the Long Sand, he ordered that the ship go about but, in the suddenly strengthening wind, he lost control of the manoeuvre. By the time he had the ship making headway again, she had drifted too far and the next thing he felt was a judder so severe that he thought the hull would disintegrate beneath his feet. They had struck the Long Sand broadside on.

The ship was under almost full sail and the grounding put huge strain on her rigging. Canvas thundered and beat above their heads, the masts buckled and jumped like fishing rods, but no one could go aloft to bring in the sails for fear of being knocked to the deck by a flailing sheet or loose spar. Instead they lit a flare and put up rockets and were relieved after a few moments to see answering signals from both the Sunk and the Knock lightships. Recalling the experience after the event, the mate of the *Indian Chief*, William Meldrum Lloyd, said: 'We could see one another's faces in the light of the big blaze, and sung out cheerily to keep our hearts up; and although we all knew that our ship was hard and fast and likely to leave her bones on that sand, we none of us reckoned upon dying.'

What they hoped was that the lightships would be able to signal to passing shipping that help was required on the Long Sand. All they could now do was wait for the dawn to come, huddled together in the deckhouses, shivering in the freezing temperatures and listening to the strengthening gale hurling massive seas against the side of the hull.

At first light a cheer rose from the ship. They had spotted the sail of what they took to be a lifeboat dodging about in the heavy seas a few hundred yards away. The men of the *Indian Chief* lined the rails, staring, unblinking, into the spray and ferocious wind, willing the boat to come closer and to brave the much larger seas which were breaking over the sands which held them so fast. But the boat, not a lifeboat as they later learned, but a fishing smack, eventually turned away, unable to help other than carry the news of the grounding back to her home port.

The tide was low when the sun came up that morning and there had been too little water to move the barque, slumped at an awkward angle on the sands. As the tide came in, her crew began to feel her shift uneasily until a larger sea rolled in, lifting her off the bottom for a second before dropping her like a brick back on the sand to the sound of splitting timbers and cargo breaking loose in the hold. The captain put out the starboard anchor in the hope that the tide might sweep the stern round so that the ship lay head to wind. But still she remained broadside on to the weather, with water pouring through the cracks opening up in her hull.

Then, with a bone-jarring crash, the ship broke her back. On the verge of panic, the crew set to work launching the three ship's boats, but each was swamped as soon as it touched the water. With no other means of escape, the desperate crew now returned to the deck cabins. Seas broke more and more frequently over the ship until, at about 5pm, a massive wave swept over the deck removing everything in its path except a few uprights of the deckhouses.

The hull was now completely awash and large parts of the deck had been blown out. The rigging was the only bleak refuge now. Some men chose to climb the foremast against the advice of the captain. He believed the mizzen was less likely to be brought down by the mainmast, which was rocking in its step and threatening to crash to the deck at any moment. Seventeen men, including the captain, his brother who was the second mate and the mate, swarmed up the mizzen mast, cutting ropes from the rigging in order to lash themselves firmly to their perch. William Lloyd, the mate, recalls:

I was next the captain in the mizentop, and near him was his brother, a stout-built, handsome young fellow, twenty-two years old, as fine a specimen of the English sailor as ever I was shipmate with. He was calling about him cheerfully, bidding us not to be down-hearted, and telling us to look sharply around for the lifeboats. He helped several of the benumbed men to lash themselves, saying encouraging things to them as he made them fast.

As darkness fell, the icy wind chiselled away at the men's will to survive. Words of encouragement between them ceased and they were left to their own feverish thoughts as the sails, ripped to shreds, thundered in the wind above their heads. In the middle of the night the mate, for a reason he could not explain, became gripped by an urge to leave the mizzen mast and join the ten men who had taken to the foremast. His only route was an aerial one via the mainmast, but he achieved it, swinging from cross-trees to stays with the strength of a desperate man.

His new refuge seemed, if anything, worse than his last. The foremast was rocking sharply as the wind tore at its sails and the mate at last began to give up hope. Then, as a massive wave broke over the hull of the ship, the mainmast toppled and fell with the sound of splintering wood and piercing cries. It had fallen aft on to the mizzen mast and brought that down with it. The mate and the other men on the foremast looked aghast at the sight of their shipmates, still lashed to the mizzen, slowly drowning as waves swept over the fallen mast, which slanted over the side into the sea.

The mate knew that it was now only a matter of time before he and the others would meet the same fate. Even the appearance a little later of a light, thought to be that of a steamer, did not encourage them much. How could anyone reach them in these conditions, even if it were a rescue attempt? But at least it gave them an incentive to hang on until dawn, whose light revealed that it was indeed a steamer they had seen, which had obviously stood by all night. The mate still could not see what help it could bring until he heard a loud cry from one of his shipmates. A lifeboat, under sail, was heading straight for them. As he later recounted:

It was a sight to make one crazy with joy, and it put the strength of ten men into every one of us. The boat had to cross the broken water to fetch us, and in my agony of mind I cried out, 'She'll never face it! She'll leave us when she sees that water!' For the sea was frightful all to windward of the sand and over it, a tremendous play of broken waters, raging with one another, and making the whole surface resemble a boiling cauldron. Yet they never swerved a hair's breadth.

Oh, sir, she was a noble boat! We could see her crew – twelve of them – sitting on the thwarts, all looking our way, motionless as carved figures, and there was not a stir among them as, in an instant, the boat leapt from the crest of a towering sea right into the monstrous broken tumble.

The much decorated Charles Fish of Ramsgate. Among his awards were two Gold Medals for bravery. (*RNLI*)

In fact, through various means, lifeboats from Aldeburgh, Clacton, Harwich and Ramsgate had been alerted to the plight of the *Indian Chief*. However, it was only the Ramsgate boat that had been able to make it to the scene of the wreck. This 42ft pulling and sailing lifeboat, the *Bradford*, had set out the afternoon before with her coxswain, Charles Fish, in command. On this occasion, his fate and that of his crew lay in the hands of one Alf Page, the master of the steam-tug *Vulcan*, which was towing the lifeboat the considerable distance she had to cover to the Long Sand into the teeth of the north-easterly gale.

Both tug and lifeboat were climbing and plunging over a heaving mountain range of sea, with clouds of water flying high over them as they struck a wave, soaking every man on board to the skin. In his account of the passage out to the wreck, Charles Fish had never encountered such a cold wind: 'It was more like a flaying machine than a natural gale of wind. The feel of it in the face was like being gnawed by a dog. I only wonder it didn't freeze the tears it fetched out of our eyes.'

A contemporary etching of the steam-tug *Vulcan*, with the Ramsgate lifeboat in tow and the stranded *Indian Chief* in the background. (*RNLI*)

Although a teetotaller himself, Fish allowed his 11 crewmen to pass around the rum they kept on board, although they were under strict instructions to leave the majority of it for those they had set out to rescue. Progress was painfully slow as the tug negotiated the ever greater seas which rampaged unrestrained, clear of the lee of North Foreland. However, by 4.30pm they had sight of the Kentish Knock lightship and drew as close as they dared, so as to get within hailing distance. They were just about able to ascertain from the lightship crew what direction to steer for the wreck, but dusk was beginning to fall now and it soon became obvious that they would never find her in darkness.

So, with extraordinary stoicism, the crew of the tug and the lifeboat agreed that their only option was to ride out the storm for the night. 'We're here to fetch the wreck,' said Bob Penny, one of the lifeboat crew, 'and fetch it we will if we wait a week.' Therefore, while the tug kept her head to sea, her paddles turning just enough to prevent her from dropping astern, the lifeboat crew did all they could to prepare for a night of extreme discomfort and anxiety. It was not a matter of getting any rest or sleep, merely of surviving a constant drenching and saving themselves from being thrown overboard as the lifeboat leapt and twisted like a hooked salmon on the end of her towline.

They did manage to rig a sort of shelter for themselves using the foresail; ten men huddled together under this while two took turns to stand in the stern, lifelines attached, to keep a lookout. Charles Fish later gave this wry account of their misery:

We all lay in a lump together for warmth, and a fine show we made, I dare say; for a cork jacket, even when a man stands upright, isn't calculated to improve his figure, and as we all of us had cork jackets on and oilskins, and many of us sea boots, you may guess what a raffle of legs and arms we showed, and what a rum heap of odds and ends we looked, as we sprawled in the bottom of the boat upon one another.

Sometimes it would be Johnny Goldsmith growling underneath that somebody was lying on his leg; and then maybe Harry Meader would bawl out that there was a man sitting on his head; and once Tom Friend swore his arm was broke; but my opinion is that it was too cold to feel inconveniences of this kind, and I believe that some among us would

The tug *Vulcan*, the steam-driven heroine of the *Indian Chief* rescue in 1881. Her crew each received the RNLI Silver Medal. (*RNLI*)

not have known if their arms and legs really had been broke, until they tried to use 'em, for the cold seemed to take away all feeling out of the blood.

Seas breaking over the boat were constantly filling the sail stretched over the men with water which pushed down on them so that some had to lie flat on their faces. When the pressure got too great, they would all heave upwards with their back to shed the water overboard. The coxswain's account continued:

'Charlie Fish,' says Tom Cooper to me, in a grave voice, 'what would some of them young gen'l'men as comes to Ramsgate in the summer, and says they'd like to go out in the lifeboat, think of this?' This made me laugh and then young Tom Cooper votes for another nipper of rum all round; and as it was drawing on for one o'clock in the morning, and some of the men were groaning with cold, and pressing themselves against the thwarts with the pain of it, I made no objection and the liquor went round.

To his dismay, when he reached into the locker for his own non-alcoholic sustenance – a bar of Fry's chocolate – Fish found that the sea had got in and reduced it to inedible pulp. There was no other food to be had on board and they were not going to try hauling themselves closer to the tug for fear of being smashed to pieces against her side.

At last the sky began to brighten in the east and before too long there was a shout; someone had seen what looked like a solitary mast protruding from a mass of white water, about 3 miles off. They soon realised that that was all that remained of the ship they had come looking for. Fish looked at the wild confusion of the sea surrounding the wreck; waves rose up in great columns as high as a house and the thunder of their self-destruction on the sands was audible even over the howling of the gale. Then he looked at his men's faces; 'let slip the tow rope', one of them urged, and with that assurance the coxswain ordered, 'up foresail!' and the lifeboat set out on her own, downwind, towards the wreck.

With all eyes fixed on the mast, the lifeboat crew were probably better off not seeing the towering seas which rose up astern of them and which spilt whole sheets of spray that flew in the wind high over their heads to land with an

Charles Fish (centre) and the crew of the Ramsgate lifeboat who helped him to save the survivors from the *Indian Chief* in January 1881. Each crewman received the Silver Medal for his part in the rescue. (*RNLI*)

explosion in their path ahead. First it looked as if no one had made it into the rigging of the wreck's remaining mast. Then, as they drew closer, the lifeboat crew realised that there was a cluster of men clinging to it but that their yellow oilskins had camouflaged them against the yellow varnished mast.

The lifeboat put out an anchor and veered in towards the remnants of the ship's stern. The men in the rigging scrambled down to the deck and picked their way aft along the rail, dodging the waves that washed across the waterlogged hull as best they could. Someone from the ship threw a piece of wood attached to a line into the water and it was grabbed by one of the lifeboat crew. Now they could make themselves fast alongside the wreck and the survivors could leap aboard.

Coxswain Fish was not prepared for the ghastly sight of all the bodies knocking about in the water among the fallen spars of the mizzen mast. He thought that the entire ship's company had been in the foremast and were therefore saved. One of those bodies, that of the second mate, was still just living and two of his shipmates shinned along the fallen mast to cut him clear and carry him to the lifeboat. Fish recalled:

> The body of the captain was lashed to the head of the mizen mast, so as to look as if he were leaning over it, his head stiff upright and his eyes watching us, and the stir of the seas made him appear to be struggling to get to us. I thought he was alive and cried to the men to hand him in, but someone said he was killed when the mizen mast fell and had been dead four or five hours.
>
> This was a dreadful shock; I never remember the like of it. I can't hardly get those fixed eyes out of my sight and I lie awake for hours of a night, and so do others of us, seeing those bodies torn by the spars and bleeding, floating in the water alongside the miserable ship.

The lifeboat's return through the tumultuous waters over the sands only added to the trauma of the 12, soon to be 11, survivors on board. They made it back to the tug and, a few hours later, to the safety of Ramsgate Harbour. Charles Fish was later awarded the RNLI Gold Medal; his crew and the entire crew of the tug, *Vulcan*, received the Silver Medal, an unprecedented number of awards for a single rescue. The mate of the *Indian Chief*, William Lloyd, paid this memorable tribute to his saviours:

> When I looked at the lifeboat's crew and thought of our situation a short while since, and our safety now, and how, to rescue us, these great-hearted men had imperilled their own lives, I was unmanned; I could not thank them, I could not trust myself to speak.

Veterans of the Ramsgate lifeboat crew who went to the rescue of the *Indian Chief* attending the naming ceremony of the town's first motor lifeboat in April 1926. They are, from left to right, Tom Friend, 81, Tom Cooper, 79, Henry Belsey, 82, Charles Vernon, 80, and David Berry, 81.

How can such devoted heroism be written of, so that every man who can read shall know how great and beautiful it is? Our own suffering came to us as part of our calling as seamen. But theirs was bravely courted and endured for the sake of their fellow-creatures. Believe me, it was a splendid piece of service; nothing grander in its way was ever done before, even by Englishmen. I am a plain seaman and can say no more about it all than this. But when I think of what must have come to us eleven men before another hour had passed, if the lifeboat crew had not run down to us, I feel like a little child, and my heart grows too full for my eyes.

6

THE *MEXICO* DISASTER, 9 DECEMBER 1886

Three lifeboats launch to the wrecked *Mexico* in a gale in the Ribble Estuary; one of them, Lytham lifeboat, rescues the entire crew and lands them safely ashore, but Southport and St Anne's lifeboats are capsized with the loss of 27 men. This, the worst disaster in RNLI history, brings about major changes in the operation of the lifeboat service and the way it is funded.

When Coxswain Thomas Clarkson stepped over the side of the brand new Lytham lifeboat, just after 3 o'clock on the morning of Friday 10 December 1886, and dropped on to the beach to lend a hand with the boat's recovery, he let out the long exhalation of an exhausted but deeply gratified man. To the cheers of a sizeable crowd, which had been waiting through the night in the wind and rain, he had just put ashore the entire 12-man crew of the German barque *Mexico*, which had run aground on a sand bank in the Ribble Estuary. In very heavy breaking seas caused by a gale from the north-west meeting a powerful ebb tide head-on, he and his crew had successfully negotiated the lethal shoals of the estuary and, in spite of a near capsize when four oars were broken, he had located the wreck in total darkness, got alongside and rescued the crew. He would later be awarded the RNLI Silver Medal for his gallant services that night.

Although the survivors were in a pretty bad way after several hours lashed to the rigging of their ship, they would all live to tell the tale and for that reason they, their rescuers and everyone ashore were exultant. Maybe the captain of the *Mexico*, G. Burmester, was more pensive than some. He had, after all, lost his ship and, to his great concern, all his ship's papers. He was also possibly playing over in his mind a very strange coincidence: only a few weeks earlier, while his ship was in Liverpool, he had visited the Liverpool International Exhibition, opened by Queen Victoria and displaying the very latest technology of the age. One exhibit that had particularly impressed him was a new RNLI lifeboat, a 37ft self-righting, 12-oared, pulling and sailing boat, fitted with the latest development in improved stability, namely water-ballast tanks running along the keel amidships. This lifeboat was none other than the *Charles Biggs*, just out of the builder's yard and destined for Lytham lifeboat station as soon

The Lytham lifeboat *Charles Biggs*, in which her coxswain, Thomas Clarkson (inset), rescued the entire crew of the *Mexico* before disaster overcame the neighbouring stations. (*Grahame Farr Archives*)

as the exhibition was over. Captain Burmester could scarcely have expected to find himself and his crew the very first people to benefit from her use, just a few weeks later on the sandbanks of the Ribble.

However bewildered he was by the night's events, the captain had not lost the art of diplomacy. The crowd on the beach had followed him and his crew to the steps of the Railway Hotel where they were to be given medical treatment and be looked after for the rest of the night; at the hotel entrance the captain turned to the people gathered and, in a thick German accent, addressed them: 'I do thank you very much, and everyone in your town, for the gallant manner in which you have this night rescued me and my crew.'

Little did he, or anyone around him cheering his words, realise the full extent of the debt the captain and his crew would actually owe the fishing communities on either side of the Ribble Estuary. Nobody knew in Lytham that two other lifeboats, from Southport and from St Anne's, had also responded to the signals of distress fired by the *Mexico* when she hit the Horse Bank off Southport. While the crowds whooped in celebration outside the Railway Hotel in Lytham, a white upturned hull glowed in the moonlight, high and dry on the sands on the estuary's southern shore, and a sombre search party picked its way among lifeless oilskin-clad bodies scattered in the vicinity. Another identical hull, also across the estuary and as yet undiscovered, was floundering in the surf, her keel also exposed to the heavens, with limp corpses, tossed about like kelp in the sea around her, entirely at the mercy of the waves.

Lytham's triumph was about to be totally eclipsed by the worst disaster in the history of the RNLI. Twenty-seven lifeboatmen lost their lives that night in a tragedy that would have a profound effect on the future of the lifeboat service.

Thanks to the inquests and investigations that followed the disaster, there are a number of recorded first-hand accounts of what happened and these, pieced together, help to tell the full story. First to speak is Captain Burmester of the *Mexico* to explain the circumstances of his shipwreck:

My ship is the *Mexico* of Hamburg. We were bound from Liverpool to Guayaquil in Ecuador with a valuable mixed cargo. The ship is iron built and barque rigged and her tonnage is 491. We left Liverpool on Sunday morning

The crew of the Lytham lifeboat outside their station soon after the *Mexico* incident, 1886. Coxswain Thomas Clarkson is in the back row, fourth from left. (*RNLI*)

(5 Dec), engaged a pilot at the bar lightship on Monday and he remained with us till Tuesday noon. We had to beat against a heavy north-north-westerly gale and our canvas was blown away.

On Thursday morning we saw the Orme's Head (by Llandudno, N. Wales). At noon we sounded 14 fathoms, at two o'clock, 10 fathoms and at four, 8 fathoms. I called my crew together and we decided to cut away the fore and mainmasts and let go both anchors. We got one anchor out at 19 fathoms and the other at 5 fathoms, but the ship continued to drift. After 105 fathoms of cable were out, she steadied for half an hour, but again began to drag. At about 9.30pm we struck the bank. We at once burned our lights and then we lashed ourselves to the remaining portion of the rigging to save ourselves from being washed overboard by the seas which were sweeping the decks.

Now Lytham's coxswain, Thomas Clarkson, takes up the story:

We set out at 10.05pm having seen the signals at 9.30pm. We steered south-west. There was a gale of wind blowing. The sea near Southport was very high and, shortly before getting to the ship, it was mountains high. Sometimes it was breaking, sometimes it was not. In its course it was running all one way. On the bank the water was properly broken. After we had gone about five miles we lost sight of the ship's lights and I said to my crew, 'Show a bright light.' They did so and it was answered immediately.

We went for about twenty minutes and lost sight of the ship's light again. As we approached, the water broke and four or five times our boat was full. I said, 'Take the masts and sails down.' As soon as that was done, the sea gave her a lurch and we broke four oars. She got partly on her beam ends. I said, 'Keep her head to the sea', and she made for the ship with her shoulder on the sea. I fancy the tide was taking us towards Southport Pier.

When we got to the *Mexico* the captain threw a black box about a foot square to the lifeboat but it went into the water. He said, 'That is the ship's statements.' I said, 'You are done, you can't get them now.' One of the men then made an attempt to get into the lifeboat and he slipped down the ship's side. He was beginning to go down and this poor fellow let go the rope just as a heavy sea was coming. It took the boat from the ship and he was knocked between the boat and the ship. Our crew pulled him in head first.

The next one came right into the middle of the boat, the next the same but the next one broke the rope and we had to get another rope. Two came down together next on one rope. The next one fell across one of the rudders and hurt his leg. The captain was next and he put the rope round his waist and the mate lowered him down. We got hold of him and swung him into the middle of the boat. I said to the captain, 'Have you seen any other boats before we came?' He said, 'No, yours is the first boat.' We could see a crowd assembled on the shore then.

These were the people of Southport who were presumably looking out anxiously for their own boat, which had indeed been launched at about 11pm. They would not have been able to see what was happening at the wreck. At about 12.15am, with all survivors on board, Coxswain Clarkson worked the Lytham lifeboat clear of the wreck, which was now heeled right over on her beam ends with massive seas washing over her decks. In spite of being full of water, the lifeboat, with all sails set, began to claw herself away from the lee shore. The coxswain takes up the story once more:

> The sea was not so heavy now. I said to the men, 'Go along the shore with the boat; show a green light and let those on the shore see that we have got the crew all right.' I said to the captain, 'Do you wish to be put ashore here or to go along with us?' He said, 'Where you go, I will go. You have a very good boat.' All in the boat were wet through and half drowned. We reached home about fifteen minutes past three o'clock. I have been in the lifeboat service 32 years. I consider the stability of the boat on which I am coxswain better than any boat I have ever sailed in.
>
> I saw nothing at all of the Southport and St Anne's boats. I knew nothing about them till a telegram came next morning stating that the Southport boat was capsized.

Then, when the news came that their neighbouring St Anne's lifeboat, *Laura Janet*, had also launched but had not returned, the same Lytham crew put immediately to sea at 10.30 on the Friday morning. Many of them had relatives aboard the missing boat and their search was therefore a desperate one. When they drew alongside Southport pier they learned the full extent of the town's loss: 13 out of 16 men aboard the lifeboat drowned (a 14th would die in hospital the next day).

Then someone on the pier spotted a white shape in the estuary near the deserted wreck of the *Mexico*. To the despair of the Lytham crew, who went immediately to investigate, it was the upturned hulk of the *Laura Janet*, partly stove in with three of her lifeless crew trapped underneath. The remaining ten bodies were subsequently found strewn along the tideline.

The sea almost claimed yet another life that day; Blackpool lifeboat had also put out when the crew heard that the St Anne's boat was missing. Her coxswain,

Robert Bickerstaffe, was swept overboard when the lifeboat, identical in design to the Southport and St Anne's boats, nearly capsized as she shipped a heavy sea. Luckily the coxswain retained a hold on the yoke line and, after being dragged some 60 yards before the sails could be lowered, he was hauled back on board.

What, then, befell the two lifeboats that were lost? No one can be certain about the St Anne's boat because all 13 men aboard her died. She was watched by many as she set off from St Anne's at 10.30 on the Thursday night, having seen the same distress signal as had the Lytham crew. Spectators watched as she was pulled away from the shore under oars until, about 100 yards out, sail was set and she disappeared into the night.

About an hour after the launch, the people who still watched and waited on the shore saw a light about a mile to windward of the one assumed to be the ship in distress. Was this, they wondered, another ship in distress or was it a lifeboat, maybe theirs, driven from her course and calling for assistance? They would never know. At about 2 o'clock they saw another light much nearer them, which they were sure was a lifeboat. It was, but it was the Lytham boat signalling her success to her own station. Surely this meant that the *Laura Janet* would soon be following? Their agonised wait lasted until 1pm on the Friday when a telegram arrived from Southport with the dreaded news that bodies from their lifeboat had been found. Exactly how the lifeboat came to grief can only be surmised. If she had capsized, she should have righted herself, unless it had been in the shallows. Maybe she struck the Horse Bank and somehow became disabled. A silver watch found on the body of one of the crew had stopped at 2.20 which suggests that the men of St Anne's struggled aboard their boat for many hours before finally succumbing to their fate.

Only two men survived the capsize of the Southport lifeboat, *Eliza Fernley*. One was Henry Robinson, one of three brothers to put out that night, and the other, John Jackson, who also had a brother aboard the boat. It is thanks to this account, given by John Jackson on the day after his ordeal, that a vivid record remains of their disastrous mission.

The barque *Mexico* was seen by a number of people at about three o'clock yesterday afternoon, riding at anchor about five miles from the spot where she came to grief. There was no reason to believe that she was in distress at the time. A gale was blowing at the time and a lookout was kept. She was kept in sight a good portion of the afternoon but as dusk crept on, we lost sight of her. The crew kept about ready to put out at a moment's notice. At last signals of distress were observed; the barque, evidently standing nearer shore, sent up rockets and flashed lights for help. Captain Hodge (the coxswain) lost no time in getting the boat out. At ten minutes to ten the horses set off with the boat and, after experiencing considerable difficulty, we launched the boat at 11 o'clock. A large crowd saw us off and the excitement was tremendous. The boat was launched successfully and went nicely for a time.

Soon, though, the sea began to assert its brutal self over the lifeboat and her crew. The oarsmen stuck grimly to their task, fighting to keep to their seat as wave after wave washed over them. Several times they were beaten back and the coxswain had to scour the darkness to find, once more, the faint glimmer of the lamp, hung from the *Mexico's* mizzen top, their only point of reference in the storm.

Eventually they drew to within 30 yards of the ship and could see that she had lost her foremast and her mainmast. Even if anyone had been shouting to them from her deck, the lifeboat crew would have heard nothing; they could scarcely hear their own voices above the wind, which was now stronger than ever. John Jackson had the anchor in his hand, ready to let it go so that the lifeboat could veer down to the *Mexico*, when a mountainous sea caught the boat amidships, picked her up and turned her upside-down. Jackson recalled:

We expected her to right herself but she remained bottom upwards. Some of us managed at length to crawl out. I and Richard Robinson held firmly onto the rowlocks and were buffeted about considerably. With some difficulty I got underneath the boat again and spoke, I think, to Henry Robinson, Thomas Jackson, Timothy Rigby and Peter Jackson. I called out, 'I think she will never right; we have all to be drowned.' I heard a voice – I think it was Henry Robinson's – say, 'I think so too.' I got out again and found Richard Robinson fairly done. He leaned heavily on my arm and I think he must have been suffocated. Another sea came and when it receded, he had disappeared and I never saw him again. While underneath I called out to my brother, 'Clasper!' – that is a sort of nickname we gave him – but I could get no answer.

The boat eventually drifted bottom upwards to the shore and those who were rescued, like myself, clung to her. I don't know what became of the rest, I was so exhausted. I remember seeing two or three struggling to reach the boat, but I do not know who they were. I drifted with the boat to the beach and staggered home, about three o'clock in the morning. I never saw any of the other lifeboats.

When the initial shock and horror of these disasters had been fully absorbed and at least some of the circumstances understood, public attention naturally turned to the organisation which supplied the lifeboats and encouraged crews to launch them in conditions such as those prevailing on that December night. In its leader on Saturday 11 December 1886, *The Times* thundered:

To praise and lament the heroism of the men who thus gave their lives is well; and some pride must mingle with the sadness caused by the story. But other

The aftermath of a shipwreck; the *Mexico* sits high and dry on the sand as salvors set to work. (*RNLI*)

reflections must arise when the circumstances are considered. What mockery, delusions and snares are lifeboats which, when upset, do not right but drift about helplessly! What a scandal to intelligence and humanity that brave men should be persuaded to risk their lives in boats so unsuitable for their work as the Southport and St Anne's must have been!

Had a tithe of the ingenuity and capital devoted to the improvement of torpedoes been expended in improving lifeboats, how different would have been their condition! Perhaps the brave crews will not have died in vain if their loss impressively directs attention to the imperfections of many of the present lifeboats, their inadequate buoyancy, their want of propelling power in face of adverse winds, and their general unfitness for much of their work. The cruel disaster may perhaps impress upon some minds the fact that the only kind of navigation in which invention has moved little, and which still depends chiefly upon men's arms, is that concerned with the saving of life.

On the following Monday, *The Times* published a terse response from the RNLI which pointed out that in the previous 32 years, self-righting lifeboats had been launched nearly 5,000 times on service and had saved upwards of 12,000 lives. Although boats had capsized on 41 of these occasions, only 18, including the *Mexico* disaster, had resulted in loss of life. In all, 76 lifeboatmen had lost their lives while on service, representing one in 850 of all the men who would have been involved.

In spite of this statistical rebuff, the criticism must have stung the RNLI. A joint Board of Trade and RNLI inquiry, while finding no major fault in the way lifeboats were operated, did expose some weaknesses. Why did the Southport lifeboat not right herself? The capability was inherent in her design and she, as with every new lifeboat, had successfully been put through the test to prove it. The conclusion was that the weight of the men hanging on to her and possibly the effect of the anchor, which had just been let go, prevented her coming upright. The feeling was that the St Anne's boat only remained upside-down because she capsized in the shallows where the self-righting action would have been hampered.

The fact that the Lytham lifeboat survived without a capsize in those conditions convinced the RNLI that improving stability in all lifeboats should be their main aim. Already 75 boats had been fitted with water ballast tanks similar to the

In 1888 a new lifeboat, the 42ft Watson-designed *Edith and Annie*, was provided for Southport and stationed at the end of the pier. (*Grahame Farr Archives*)

Lytham boat. That work would now be accelerated. After the disaster, Southport and St Anne's received sailing lifeboats with drop-keels, to be moored at the end of the pier and to supplement their carriage-launched boats.

The Times seemed to suggest in its leader that it was primitive in the late 19th century for lifeboats still to rely solely on muscle power. It is likely that the RNLI's development of steam-driven lifeboats was given new impetus by this event, although the first one did not appear until 1891. The concept was not practicable in most locations, however, and it would not be until the late 1910s that crews began to put their faith in the internal combustion engine as a substitute to their arms.

Looking at the incident with 21st-century eyes, it seems extraordinary that the inquiry was not troubled by the lack of coordination between the lifeboat stations involved. It is probable that the Lytham boat had rescued all the men

The townspeople of St Anne's show their support when their new lifeboat, *Nora Royds*, is named and ceremonially launched. (*RNLI*)

from the *Mexico* before either of the other two lifeboats had even got close to the scene. But in those days, when the only known communication at night in a storm was a light or a flare, there was little chance of one lifeboat knowing another's actions or distress, even if they knew they had launched.

And what of the men who lost their lives? Was there more that they could have done for their own safety? The inquiry found little to fault at Southport, and although some of the crew had apparently remonstrated with their coxswain for keeping the boat too often broadside to the sea, this was seen as something almost impossible to avoid in such conditions.

At St Anne's there was more to be concerned about. It seems that the coxswain of the lifeboat, William Johnson, a 35-year-old fisherman, was very sick with consumption and had not been expected to live longer than a few months. Two or three of the other crewmembers were also not strong men, one of them having only had 'a basin of gruel all day', apparently stinting himself on behalf of his wife and children. The station's honorary secretary was unaware that the lifeboat was being launched, so there had been no one in authority to question the fitness of the crew to put out that night.

It is important to remember that in those days, to a poor fishing community, the few shillings to be earned by volunteering to take an oar of the lifeboat was a considerable bonus. A man who was not eating so that his children could

The dedication ceremony of the memorial to the 13 men who died aboard the St Anne's lifeboat. (*RNLI*)

would also have the greatest incentive to hurry to the lifeboat when a distress signal went up. The need to ensure that the decision to launch was made by the honorary secretary who remained ashore became very obvious to the RNLI following the inquiry.

To a Victorian public, whose every newspaper was filled with the account of 27 selfless, brave, impoverished lifeboat volunteers sacrificing themselves in a winter storm and leaving widows and scores of children bereft and penniless at Christmas time, this was a tragedy on a scale that even Charles Dickens would not have dared to invent. Their response was immediate; a fund to benefit the families was established and donations poured in so that it finally realised some £50,000.

One of the earliest 'Lifeboat Saturday' street collections held in Manchester in the 1890s. It was organised by Sir Charles Macara of St Anne's. (*Weekend*)

One man was pivotal, not just in the establishment of the widows' fund, but in harnessing the wave of sympathy and admiration for lifeboat crews that swept the nation for the lasting good of the RNLI. Charles Macara, a Manchester businessman, owned a bungalow among the houses that were springing up in the new resort of St Anne's. He was a member of the St Anne's Lifeboat Committee and when anxious relatives waited for news on the morning of 10 December they gathered round his house as he was the only man in the town to own a telephone. Sadly, the enquiring telegrams he ordered to be sent up and down the coast all came back with negative replies.

After the disaster Macara wanted to do more for the RNLI and decided that the Institution's income was dangerously dependent upon a wealthy but narrow group of subscribers. His idea was to bring the appeal of the lifeboat service to the man in the street – quite literally – by parading an RNLI lifeboat through the streets of Manchester and asking shoppers and onlookers to place their silver and copper donations in purses attached to long poles. The first of these 'Lifeboat Saturdays' took place in 1891 and the idea soon caught on in other large cities throughout Britain. This method of collecting soon evolved into the Lifeboat Flag Day that we know today and is believed by many to have been the original inspiration for all charity street fundraising.

Fraserburgh lifeboat, *Duchess of Kent*, photographed standing by the Danish fishing vessel she had put out to assist moments before her capsize in January 1970.

7

FRASERBURGH, 30 JUNE AND 7 SEPTEMBER 1909

Andrew Noble, coxswain of Fraserburgh's pulling lifeboat, earns the Silver Medal twice in the same year for rescues to herring drifters at the harbour mouth.

At the time of writing, Charlie Duthie is the lifeboat operations manager at Fraserburgh lifeboat station. He is the man who decides whether to send the lifeboat out through the east-facing harbour entrance when a call comes, often when the weather and sea are at their most treacherous outside this fishing port on the exposed north-eastern shoulder of the Aberdeenshire coast.

With the powerful, high-tech Trent class lifeboat at his crew's disposal, the decision to launch is not usually as difficult to make as it used to be for Duthie's predecessors. In the early days of motor power, they had to weigh up the chances of a single 40hp engine propelling the lifeboat safely through the harbour entrance with waves sweeping across it in a north-easterly gale. Earlier still, and it was only the power of 10 or 12 men at the oars that stood between a successful rescue and disaster.

Tragically, the town of Fraserburgh has encountered more than its fair share of lifeboat disasters. Early on the gale-swept morning of 21 January 1970, Duthie, then a regular member of the crew, arrived breathless at the station to find that the lifeboat had launched moments earlier without him. His frustration at missing the shout turned later that day to a feeling of bemused horror when he heard that the lifeboat, a 46ft 9in Watson class, *Duchess of Kent*, had been

An angry sea awaits Fraserburgh lifeboat, *Duchess of Kent*, as she puts out through the harbour entrance in October 1959. (*George M. Wilson*)

pitch-poled, bow over stern, by a gigantic wave while attending to a Danish fishing vessel 40 miles out into the North Sea. Five of her six-man crew drowned underneath the non-self-righting boat; the sixth, John Jackson Buchan, was thrown clear and was saved, clinging to the up-turned hull, by another vessel that had been standing by.

Seventeen years before that, on 9 February 1953, six of the seven men aboard the *Duchess of Kent*'s predecessor, *John and Charles Kennedy*, had also been drowned when the lifeboat capsized at the entrance to Fraserburgh Harbour while escorting fishing boats to safety. So much of the important business of Fraserburgh lifeboat was conducted at the harbour mouth throughout the station's history. For fishermen returning with their catch, the lights of Fraserburgh might have been a welcome sight but they also knew in bad weather that they were approaching the most perilous moment of their expedition. For Andrew Noble, the town's most decorated coxswain, the harbour entrance was the scene of his greatest triumphs but also his ultimate tragic demise.

Andrew Noble took over the position of coxswain in 1887. In his time as coxswain he would see the transition from oar and sail power to motorised lifeboats, but it was while still at the helm of a pulling boat in 1909 that, twice within the same year, he earned the Silver Medal for bravery in full view of the townspeople of Fraserburgh.

The first occasion was in mid-summer when the southerly track of the migrating herring shoals had reached the waters off Scotland's east coast. The lucrative herring industry was at its peak in the early 1900s and vast fleets of drifters were constantly at work between the fishing grounds and the harbours of Arbroath, Montrose, Aberdeen, Peterhead, Fraserburgh, Macduff, Banff, Buckie and several more.

A summer gale sprang up from the north-east on the morning of Wednesday 30 June 1909 and, much to the concern of everyone ashore, it was sending powerful breakers directly across the harbour entrance at Fraserburgh. The brown sails of the approaching and heavily laden herring fleet were in sight. Two hundred boats were about to run the gauntlet across the line of rampaging seas. No one was surprised to see the lifeboat launch and take up a position in the channel

The *John and Charles Kennedy*, wrecked against the outside of Fraserburgh's south harbour wall after she had capsized escorting fishing boats through the harbour entrance in February 1953. Six of her seven-man crew were drowned.

A crane is used to right the *John and Charles Kennedy* after the 1953 disaster off Fraserburgh and, opposite, water is pumped from the wreckage. (*Both photographs: Geo. A. Day*)

ready for any emergency, as the drifters drew close to the pier heads. The men of the rocket brigade were also setting up their lifesaving equipment on the South Breakwater.

A sizeable crowd was forming on the opposite pier to watch with breathless anxiety as each boat made its run for safety. While some made it through in a relative lull between waves, others were lifted high on to the crest of a roller, carried out of their course and spun round so as to be heading straight for the

Andrew Noble, coxswain of Fraserburgh's pulling lifeboat, earned the RNLI's Silver Medal twice in 1909. (*RNLI*)

South Breakwater. Drastic avoiding action saved most from destruction, although several suffered broken mizzen masts and the *Heatherbell* of Rosehearty had all her nets washed away as she fell from the top of a wave, missing the breakwater by inches.

The *Henry and Elizabeth* of Nairn was not so fortunate. Just as her master, William Barron, was lining her up for the entrance, a towering breaker smashed into her side, enveloped her and left her lying on her beam ends, her mainsail submerged. Six of the seven people on board had been able to clutch hold of something to keep them in the boat. One man, though, Alexander McIntosh, was in the water. None of his shipmates had seen him go but the crowd on the breakwater had and they were shouting for someone to save him. He fought like a fiend to keep afloat as waves broke over him but then, quite suddenly, he threw up his hands in despair and disappeared beneath the surface, never to reappear.

Meanwhile, the five men and the boy still aboard the capsized fishing boat were relieved to feel their vessel heave back upright after her mainsail burst, releasing the water that had been holding her down on her side. The drifter was, however, now powerless to save herself. Her crew, realising at last that one of their number was missing, clung helplessly to their vessel as she drifted along the outside of the South Breakwater, bumping over the rocks near the retaining wall of the reclaimed ground.

Coxswain Andrew Noble, standing in the stern of the 12-oared, self-righting lifeboat *Anna Maria Lee*, would have seen the knock-down and heard the shouts from the people on the pier. By the time his oarsmen had brought the lifeboat close to the stricken vessel, the man in the water had vanished and it was the men still aboard the now grounded fishing boat who were his main concern.

June 1909: with a line aboard the stricken herring boat *Henry and Elizabeth*, Andrew Noble and his crew haul themselves close enough to allow survivors to scramble aboard the lifeboat.

Manoeuvring round the lee side of the *Henry and Elizabeth*, the lifeboat crew got close enough to put a line aboard. This gave them the means to haul themselves alongside the casualty long enough for one man to leap between the two vessels. Although aground, the herring boat would occasionally be lifted high above the lifeboat before falling again on to the rocks and threatening to crush the lifeboat as she drew close.

Somewhere in the chaos, the yoke of the lifeboat's rudder was destroyed so that the coxswain could no longer control it from inside the boat. Instead, with the six survivors aboard, the lifeboat had to work her way out of the surf with the coxswain at full stretch over the side of the boat, manipulating the rudder with his bare hands. Spectators on the pier held their breath as they watched the lifeboat lifting nearly vertically on the crest of a wave, half her length clear of the water, before her bow crashed down into the following trough. Each of these convulsions weakened Andrew Noble's grip on the rudder and every passing wave threatened to wash him over the side. But the men at the oars kept stubbornly to their task and, yard by yard, the lifeboat drew closer to the harbour entrance.

A huge cheer rang out from the people gathered on the breakwater when they could see that the lifeboat was suddenly clear of the breakers and under the shelter of the tall harbour wall. From the events they had just witnessed it was an immense relief that more vessels and more lives had not been lost. That relief would, of course, have been little comfort to the widow and four children of the unfortunate Alexander McIntosh.

Coxswain Andrew Noble and his crew won many plaudits. One eyewitness, a Captain F.J. Noble of the Fraserburgh Harbour Board, made this comment:

Well, I confess, I've seen some seamanship in my time, in all parts of the world but I never seen with my two eyes a boat handled like yon in my life. When I saw the lifeboat make for the back o' the South Breakwater where the breakers were raging among the rocky shallows, I tell you honest, I thought it was madness. I stood there stunned and waited for the moment of destruction. But nothing could have defeated yon brave fellows. Talk about cool courage. They picked their way to the wreck as dainty as a lady when she goes shoppin'. The rescue was grand and the return to the harbour without even scratching the paint, a miracle. Andrew Noble has a fine record as coxswain of the lifeboat, but this last feat, in my humble opinion, is his greatest.

The Committee of Management of the RNLI must have agreed whole-heartedly with such an assessment when they decided to confer the Silver Medal on Andrew Noble at their meeting in London on 12 August 1909. Little did they expect that, only two months later, they would be examining yet another report and recommendation to recognise Coxswain Noble's bravery after a second rescue in very similar circumstances.

Another unseasonal gale, this time from the north-west, had grown into a fury by the morning of Tuesday 7 September, sending towering breaking seas crashing into the North Breakwater of Fraserburgh Harbour and making the entrance a life or death lottery to negotiate.

This did not deter a small number of fishing boats that needed to get out of the ferocious weather to land their catch. At lunchtime on that day only a few onlookers were present to see a steam drifter loom out of the haze that enveloped the bay and make her run for shelter. She happened upon a lull at the bar and made it into sheltered waters with little difficulty. Then the shape of a close-reefed sailing boat, the yawl *Zodiac* of Buckie, appeared. Three massive waves

September 1909: Fraserburgh lifeboat, *Anna Maria Lee*, approaches the yawl, *Zodiac*, hard aground beneath the south wall of the harbour.

caught her on the broadside in quick succession and swept her past the entrance. It looked as though she would be carried up among the rocks on the south side, but she recovered enough way to allow her back in between the pier heads.

All seemed under control as the tugboat, *Hugh Bourne,* held in readiness in the channel, moved towards the fishing boat to pass a line and tow her to safety. But just at that moment a succession of thunderous breakers swept across the entrance and carried the yawl helplessly up the back of the South Breakwater and on to the same outcrop of rocks on which the *Henry and Elizabeth* had foundered three months earlier.

Almost immediately, loud explosions rang out over the town. The signals to summon both the lifesaving appliance company and the lifeboat crew had been fired, but their effect had also been to alert what seemed to be the entire population of Fraserburgh which now, disregarding the ferocious wind, swarmed towards the harbour for a view of any drama that might be about to unfold.

Chief Officer Glanton of the lifesaving company had soon organised his men on the South Breakwater and with a whoosh and a billow of smoke, a rocket with line attached was unleashed. It caught in the foremast of the fishing boat whose crew quickly grabbed the line and began to haul in the heavier line and block that would allow the breeches buoy apparatus to be rigged.

Just as the youngest member of the seven-man crew had clambered into the breeches buoy, ready to be hauled up on to the breakwater, the lifeboat appeared, her 12 oarsmen straining every muscle as she rounded the southern pier head. Seeing the lifeboat now surfing towards him, the skipper of the stricken *Zodiac*, Peter Coull, changed his mind and ordered the boy out of the breeches buoy; he preferred to put his faith in a boat-to-boat transfer rather than one where his crew would be suspended over the rocks and breakers underneath the pier.

Andrew Noble, at the helm of the lifeboat, using his considerable skill and the experience gained during his June rescue, steered the lifeboat alongside the yawl and held her there long enough for all seven fishermen to clamber aboard the *Anna Maria Lee*. As the lifeboat turned for home, the coxswain and his crew soon realised that they had a well-nigh impossible struggle on their hands. The tide was low and in the shallow broken water they were making practically no progress into the wind. The thousands of people now lining the pier watched as waves broke continually over the lifeboat, drenching her occupants and sapping their strength. If anything, she was getting nearer the rocks and many feared they were about to witness a disaster.

Then there seemed to come a pause as if the sea needed to regain its strength. The lifeboat crew understood that this was their chance for one last desperate heave on the oars to pull themselves into more open water. Everyone on board, survivors and crewmen alike, bent to the task and gradually they drew clear of the shallows where the sea was less troubled and an anchor could be put over the side.

There was hope now that the tug *Hugh Bourne*, which had been dodging about near the harbour entrance, could get to them. As she left the shelter of the piers, huge seas swept over her and quantities of steam could be seen pouring from every aperture. It was soon obvious that her engine rooms had been filled with water and her fires were all but out. With what little power she had left, she turned back and made it safely into the harbour.

The lifeboat, still at anchor about half a mile out in the bay, often disappeared from view in a deep trough before re-emerging teetering on the white crest of a wave. The coxswain soon realised that he would have to get home without the help of a tug and so ordered the sail to be hoisted. He then headed out to the east with a view to running down with the wind to the harbour mouth. As he looked back towards the town, he was astonished to see the unmistakable shape of a steam drifter, its long, thin funnel wafting white smoke like incense as it swung wildly with the vessel's uneasy progress towards him. She was the *Lively* of Buckie and her skipper, Alexander Thomson, had been getting up steam inside Fraserburgh Harbour just in case other help for the lifeboat failed.

He had made it safely through the harbour entrance and before long he was alongside the lifeboat and had passed a towline into the grateful hands of an exhausted crew. The crowd on the harbour walls gave a heartfelt cheer as the two vessels reached safety. Not only had they witnessed the rescue of a sailing fishing crew by Andrew Noble and his men, but they had also seen the compliment courageously returned by the skilful and unsolicited assistance given by a powered steam drifter to the pulling lifeboat.

Although all human life was safe, the crew of the *Zodiac* remembered, too late, that in their haste to board the lifeboat they had forgotten about the little dog they always took to sea with them. A young man spotted it from the pier as it leapt from the swamped vessel and swam towards the rocks. The young man, risking being swept away by the backwash, plunged through the surf and brought the bedraggled animal safely ashore.

Although there is no obvious record of any official recognition for this canine rescue, the deeds of Captain Alexander Thomson of the *Lively* and of Coxswain Noble were justly rewarded. The very next day the Harbour Commissioners made a public presentation of £25 to the drifter skipper and the RNLI later presented their official Thanks of the Institution on Vellum to him and awarded a second Silver Service Clasp to Andrew Noble.

Nearly ten years later and Andrew Noble was still coxswain of the lifeboat, having held the position for 32 years. On 28 April 1919 he set out with his 12-man crew in a severe north-north-easterly gale aboard the station's new motor powered lifeboat, *Lady Rothes*. The Admiralty drifter *Eminent* was floundering with a failed engine not far from the harbour entrance. Close to the casualty, the lifeboat was hit by two huge waves and capsized. Only three men found themselves still aboard when the lifeboat righted herself. A few managed to regain the boat, but the others were swept with the powerless lifeboat towards the nearby beach. Coxswain Noble, one of those in the water, shouted to those aboard the boat, 'Let go the anchor!'

When rescuers reached the beach they found the lifeboat washed up and 11 men alive. There were also two lifeless corpses. One was Acting Second Coxswain Andrew Farquhar, and the other Coxswain Andrew Noble.

The crew of the *Eminent* survived thanks to the rocket apparatus team after the drifter also ran aground on the beach.

1. Courage and determination have been a constant characteristic of lifeboat crews, even if the skills required today are different. (*RNLI*)

2. The artist Tim Thompson's depiction of Ramsgate Coxswain Charles Fish and his crew sailing away from the wreck of the *Indian Chief* in January 1881 with the survivors on board. Many of the stricken ship's crew were lost during the two nights they were aground and the lifeboat crew had to survive a whole night in tumultuous seas before they could carry out the rescue (see Chapter 5). (*Painting by Tim Thompson*)

3. William Wyllie's depiction of Southsea lifeboat attending a wreck. The perils of getting alongside a ship in a storm are as great today as they were in the days of sail and oar. (*RNLI*)

4. *Return of the Lifeboat to the Quay* by William H. Barrow. Lifeboats in rough seas made a dramatic subject for marine artists of the Victorian era. Their work helped to bring the realities of rescue at sea to inland audiences and to build support for the RNLI. (*RNLI*)

5. *A Shoreboat Service* by A.E. Emslie (1888). In its early days, the RNLI spent as much effort encouraging and rewarding rescues by people using their own boats as it did in providing dedicated lifeboats. Outstanding shoreboat rescues are still recognised by the Institution – this painting hangs at its headquarters in Poole, Dorset. (*RNLI*)

6. In the days when men had to pull against tide and weather to reach a wreck, they sometimes stood a better chance if the lifeboat was transported overland, often for several miles, to a more favourable launching point. (*RNLI*)

7. The image of Sir William Hillary appears on the Gold, Silver and Bronze Medals awarded by the RNLI (see Chapter 1). (*RNLI*)

8. and 9. By the 1920s and 1930s, Lifeboat Day collectors had become a familiar, if eccentric, sight on the streets of London. Before Sir Charles Macara's movement in the 1890s to bring support for the RNLI from the man on the street, the Institution relied on only a narrow group of wealthy benefactors (see Chapter 6). (*Keystone View Co*)

10. Tim Thompson's painting of the moment in February 1936 when Ballycotton lifeboat crew hailed the Daunt Rock lightship to discover her intentions. The Ballycotton crew spent 49 hours at sea during this marathon rescue, which ended with the lightship crew being rescued after their vessel had been driven close to the rocks at the entrance to Cork Harbour (see Chapter 8). (*Painting by Tim Thompson*)

11. The memorial erected in Southport to commemorate the 14 men who were lost after their lifeboat overturned in mountainous seas just 30 yards from the stricken *Mexico*, 9 December 1886 (see Chapter 6). (*Grahame Farr Archives*)

12. Henry Blogg, coxswain of Cromer lifeboat for 38 years in the early 20th century, gained the status of a national hero and is here captured in oils by the celebrated contemporary portraitist Thomas Dugdale (see Chapter 9). (*Painting by Thomas Dugdale*)

13. Cromer's No. 1 lifeboat, *H.F. Bailey*, used in the rescue of 88 men from a wartime convoy that ran aground on Haisborough Sands in August 1941, is on public display in the Henry Blogg Museum on the seafront at Cromer (see Chapter 9). (*Charter Consultant Architects*)

14. (Opposite) The statue of Richard Evans, which overlooks the coast at Moelfre, was sculpted by Sam Holland and unveiled in 2004 by HRH The Prince of Wales. Evans was awarded a Gold Medal for his part in the rescue of 15 men from a freighter in hurricane-force winds in December 1966 (see Chapter 11). (*RNLI*)

15. (Above) The memory of Richard (Dick) Evans, BEM, coxswain of Moelfre lifeboat, is preserved in oil by Jeff Stultiens, whose other subjects include Queen Elizabeth II and Cardinal Basil Hume. After his retirement in 1970 Evans used his skills as a natural orator to recount his adventures at various functions and was invaluable to the RNLI in public relations (see Chapter 11). (*Painting by Jeff Stultiens*)

16. The moment in December 1981, before the Penlee lifeboat, *Solomon Browne*, made her second and fateful run in alongside the stricken *Union Star*, as interpreted by the marine artist Tim Thompson. No one saw what happened next and no one aboard either the ship or the lifeboat survived the catastrophic consequence (see Chapter 13). (*Painting by Tim Thompson*)

17. The remote settlement at the tip of Spurn Head, home for the Humber lifeboatmen (see Chapters 10 and 12). (*Les Stubbs*)

Idea Store Bow

Items that you have returned

Title: Lifeboat heroes : outstanding RNLI
rescues from three centuries
ID: 91000001537620

Total items: 1
10/02/2023 15:50

24 hour renewal line: 0333 370 4700
www.ideastore.co.uk

18. The 47ft Watson class lifeboat *Solomon Browne* puts in an appearance on a harbour day in Mousehole, Cornwall (see Chapter 13). (*Lalouette Photographs*)

19. In 1983 Penlee received a new lifeboat, the Arun c̄ [...] ꓛmon *Browne* and her crew of eight men from Mousehole ir [...] ed to ensure that crews were given the best possible tools fo [...] rnise the fleet (see Chapter 13). (*Frank Austin*)

20. Atlantic 21 crew Rod James (left) with Christopher Reed (centre) and Warren Hales, who rescued survivors from the sail-training ketch *Donald Searle* off Hayling Island, October 1992 (see Chapter 14). *(RNLI)*

21. The Atlantic 21, *Aldershot*, in which Rod James performed his two Silver Medal rescues from Hayling Island in 1981 and 1992 (see Chapter 14). *(RNLI)*

22. The Atlantic 75, a rigid inflatable lifeboat now stationed at Hayling Island for inshore rescues, turns to face a breaking sea head-on (see Chapter 14). (*Rick Tomlinson*)

23. The Atlantic 75, *Leslie Tranmer*, stationed at Southwold in Suffolk, at her full 35-knot speed. (*Nicholas Leach*)

24. Coxswain Hewitt Clark, MBE, the most decorated lifeboatman of his time, aboard the Lerwick lifeboat with his son, Neil, who later became the coxswain. Hewitt was awarded the Gold Medal for his part in the rescue of 15 men from the freighter *Green Lily* off the Shetland island of Bressay in November 1997 (see Chapter 15). (*Edward Wake-Walker*)

25. The Severn class lifeboat *Michael and Jane Vernon*, which, with Coxswain Hewitt Clark and his crew from Lerwick, was involved in saving the crew from the *Green Lily* in 1997 (see Chapter 15). (*RNLI*)

26. Today's lifeboat crews may be equipped with the most up-to-date vessels and every technological aid, but the conditions they face at sea are no easier than those experienced by the first lifeboat crews two centuries ago. (*Rick Tomlinson*)

27. Scarborough's 14-tonne Mersey class became airborne on several occasions during her search for a surfer in Force 6 winds and a 15ft swell in November 2007 (see Chapter 16). (*Edward Behan*)

28. Winter seas off Buckie, Grampian, provide a gruelling test for the town's Severn class lifeboat, *William Blannin*, and her six-man crew. (*Nigel Millard*)

29. A Coastguard helicopter of the type that lifted 12 men off the MV *Ice Prince* in January 2008, with the Yarmouth, Isle of Wight, Severn class lifeboat, similar to the one from Torbay which rescued eight of the survivors (see Chapter 16). (*M.B. Angus*)

8

BALLYCOTTON,
11 FEBRUARY 1936

Coxswain Patrick Sliney is awarded the RNLI Gold Medal after 'one of the most exhausting and courageous rescues in the history of the lifeboat service'. He and his crew spend 49 hours at sea in a gale and bitter cold to save the crew of the Daunt Rock lightship close to the rocks at the entrance to Cork Harbour.

Until its decommissioning in 1974, the Daunt Rock lightship, marking a lethal hazard near the mouth of Cork Harbour, was a familiar and often reassuring sight to Atlantic seafarers returning to the Irish mainland. For the tens of thousands of emigrants sailing out of Cobh (or Queenstown as it was previously known), the red-painted vessel was one of the last symbols they would glimpse of their old country, their old life, before surrendering themselves to the turmoil of an ocean crossing and the uncertainty of a new continent.

With modern aids such as radar and satellite navigation, only a buoy is required today to mark the same hazard, but for 100 years men were paid to keep the light ablaze in all weathers, 365 days a year. Lightships have no means of propulsion, so their only safeguard in times of high winds and heavy seas is their anchor and cable. Both are substantial; the mushroom-shaped anchors commonly used weigh some 6,000lb (2,722kg) and each link of the 450ft chain weighs 14lb. With the Daunt Rock lightship's position only 2½ miles from a rocky shore, her crew were doubtless often grateful for every ounce of strength in that cable.

In October 1896 the entire eight-man crew of the Daunt Rock lightship, *Puffin*, was lost when the vessel disappeared in a gale. Whether she was blown on to rocks or overwhelmed by the sea was never certain, but a study of her wreckage recovered from the sea bed suggested that the mast carrying the light had been torn down and the iron and wood composite vessel was filled with water through the gaping hole left in her deck.

Puffin	*Comet*
Built 1886/87 by Schlesinger Davis & Co, Wallsend; length 91ft, breadth 21ft, depth 11¼ft; construction composite; cost £6,000; sank during storm on Daunt, 8 October 1896, crew of eight lost. Salvaged by Ensor & Sons. Beached at Rushbrooke, sold 27 October 1897 to Ensor and scrapped on beach.	Built 1904 by J. Reid, Glasgow; length 96ft, breadth 23ft, depth 12¼ft; construction iron shell and floors, steel framing; five watertight bulkheads; steel mast and fixed lantern; mizzen mast carrying day mark; cost £6,740; sold in 1965 to Turner & Hickman, Glasgow (Shipbrokers), and subsequently used as a broadcasting station by Radio Scotland.

Forty years on and the *Puffin's* sturdier, iron and steel replacement, *Comet*, was feeling the effects of a February cyclone that had wound itself up in the Atlantic and which, by Monday 10 February 1936, was unleashing a hurricane-force onslaught from the south-east, straight at the Cork coast. Memories of the disaster to the *Puffin* were still vivid in the minds of many in the port of Cobh and when news reached them that distress rockets had been seen coming from the lightship early on the Tuesday morning and that she was no longer on her station, a rumour swept through the town that yet another crew had been lost with all hands. Two of the men were from Cobh and their families must have felt immense relief to hear a little later that she had not yet gone ashore, but had drifted about half a mile from her station. She was, however, perilously close to the rocks off Robert's Cove. One of her crew later described their predicament:

You cannot imagine, unless you went through the same experience yourselves, the awful effect these rocks had upon us since we dragged our anchor and we had drifted to within half a mile of their reach. They seemed to be continually grinning, defying our efforts to thwart the relentless fury of the sea.

On Monday night the wind went mad in its fury, the seas continually pounded us and, indeed, it looked as if we were lost. At one o'clock on Tuesday morning our cable – of which we had about 190 fathoms – parted and we were being blown towards those hated rocks of Robert's Cove. We worked like demons to run out our second cable. Our first attempt was a failure, for it fouled; but we had enough compressed air to bring it inboard and free it for the second attempt. Those 20 minutes taken to perform this operation seemed like an eternity. We made another attempt and you can guess the joy we experienced as we felt the anchor grip the bottom and check our progress to destruction.

Then one of the crew tapped out the SOS and this was picked up by the German tug, *Seefalke*, as she battled onwards through the seas 37 miles away to render assistance to the crippled British freighter, *Baron Graham*, also in danger of being dashed ashore to her doom on the Waterford coast. The tug communicated our danger to the pilot cutter on duty in Cork Harbour, an act for which we all would like to express our grateful appreciation.

You can imagine with what relief we watched the Cork–Fishguard passenger steamer, *Innisfallen*, approach us at 8.30am that morning and also at the arrival of the British destroyer, HMS *Tenedos*, from Cobh.

That SOS message from the lightship was to set in motion one of the most epic services ever performed by an RNLI lifeboat crew, one that not only proved the all-weather capabilities of the latest type of motor lifeboat, but the astonishing endurance of the men of a small southern Irish fishing village.

The effects of the hurricane on the fishing harbour at Ballycotton, which lies 10 miles to the east of the entrance to Cork Harbour, were unprecedented. In the early hours of the Monday morning, gigantic waves were exploding against the harbour wall with such force that hefty stone blocks were being torn from their seating and tossed about on the quay. Patrick Sliney, a seasoned pillar of the local fishing community and coxswain of the lifeboat said later that he and other men had spent 'a night of terror' trying to save the boats from destruction inside the harbour. Little did he know that his nightmare had only just begun.

His story and that of his crew, which included his brothers Tom and John and his son William, was graphically recounted by the Ballycotton lifeboat honorary secretary of the day, Robert Mahony, who sent his report to RNLI headquarters in London a few days later:

During the Sunday and early on Monday the coxswain ran ropes from the lifeboat, the *Mary Stanford*, a 51 feet Barnett cabin motor lifeboat, to prevent her from striking the breakwater. At midnight on the Monday, when the gale had risen to a hurricane, the coxswain's own motorboat was seen to have parted her moorings, and was in danger of being carried out to sea. The coxswain and several other men attempted to launch a boat to her, but were nearly swamped. Stones, a ton in weight, were being torn from the quay and flung about like sugar lumps. I spent most of the night near the lifeboathouse, watching the terrible destruction that the wind and waves were doing. Twice I was spun round and nearly flung on my face. At three on the Tuesday morning I went to bed, but not to sleep. I was out again shortly after seven, and found that the coxswain and the other men had been up all night trying to secure his motorboat. They had succeeded in launching a boat, got a rope to the motorboat and secured her. It was at that moment, after this long night of anxiety, that the call for the lifeboat came.

With the telephone lines torn down by the gale, a messenger had carried the news in person from Cobh to Ballycotton that eight men were adrift aboard the Daunt Rock lightship, 12 miles to the west. It was soon after 8 o'clock in the morning when the honorary secretary passed the news to his coxswain who had just got home. Patrick Sliney made no reply. Both men had seen the seas breaking over the lifeboathouse where the boarding boat was kept and Robert Mahony was convinced there was no way of getting aboard the lifeboat at her mooring, let alone leaving the harbour. He could not order a launch in such weather.

Sliney, ignoring his exhaustion from the night's battle with the moorings, nevertheless decided to go back to the harbour to see if anything could be done. His honorary secretary followed him a few minutes later and was astonished to see the lifeboat already at the harbour mouth and dashing out between the piers. As he later recounted:

The coxswain had not waited for orders. His crew were already at the harbour. He had not fired the maroons, for he did not want to alarm the village. Without a word they had slipped away. As I watched the lifeboat I thought every minute she must turn back. At one moment a sea crashed on her; at the next she was standing on her heel. But she went on. People watching her left the quay to go to the church to pray.

Two small islands lie off Ballycotton, the further of which, a mile off the coast, supports a lighthouse. The lifeboat had reached the outer island and was now meeting seas of such enormity that their spray was flying clean over the 196ft lantern of the lighthouse. The lifeboat seemed to hesitate and then turned round. Was she coming back? No; her coxswain had decided rather to take her through the sound between the two islands. This was far more risky than the open sea route, but he would save half a mile and perhaps reach the lightship before it was too late.

The seas in the sound were terrifying for the crew. Every one of them was in the after cockpit and each time a wave swept through it, Patrick Sliney counted his men. At one point the lifeboat toppled from the crest of a massive sea and fell with such a thud into the trough beyond that all on board were convinced that the engines had gone through the bottom of the boat. All waited anxiously while the motor mechanic, Thomas Sliney, went to check on the damage. 'All's well,' he was able to report to his brother, the coxswain. 'After that, she'll go through anything.'

Safely through the sound and now about 6 miles from Ballycotton, Coxswain Sliney found the following seas were worse still. During the process of putting out a drogue, waves continually swept over the lifeboat, half-stunning the coxswain as they crashed over his head. The largest wave filled the cockpit and knocked every crewman off his feet.

Visibility was appalling in the spray and sleet. The lightship could not be found and finally the coxswain decided to run for Cobh for information.

The pilots at Cobh were able to give an exact position; the lifeboat set out again and soon after midday found the lightship a quarter of a mile south-west of the Daunt Rock and only half a mile from the shore. Her crew would not leave her, knowing the danger an abandoned lightvessel out of position would present to shipping. They feared their anchor would not hold and asked the lifeboat to stand by.

The——
Royal National Life-boat Institution.

(Supported solely by Voluntary Contributions.)

IRISH FREE STATE DISTRICT.

PATRONS:
His Excellency the Governor General.
PRESIDENT OF THE LADIES' LIFE-BOAT GUILD.
Her Excellency Mrs. James McNeill.

INAUGURAL CEREMONY
OF THE
BALLYCOTTON MOTOR LIFE-BOAT

" MARY STANFORD."

at

BALLYCOTTON HARBOUR,

On TUESDAY, JULY 7th, at 10.15 a.m.

The Life-boat to be named by
MRS. WILLIAM T. COSGRAVE.

The cost of this Life-boat, £11,000, has been defrayed out of
a legacy from the late Mr. J. F. STANFORD, of London.

PRICE THREEPENCE.

Above and opposite: The front and back cover of the programme for the naming of
the Ballycotton lifeboat, *Mary Stanford*, on 7 July 1930. *(RNLI)*

The Life-boat Service in the Irish Free State

There are

18 Life-boats

on the coasts of the

IRISH FREE STATE,

of which

14 are powerful Motor Boats,

ready at any time of the day or night to answer the call of the shipwrecked.

Another powerful Motor Boat is under construction.

Last year the Irish Life-boats were called out on service 21 times, and 8 of these launches were to the help of Irish vessels.

£10,000 a year is spent

on maintaining the Life-boats in Ireland, exclusive of the capital expenditure on new boats ;

£5,885 was contributed by Ireland last year.

The Institution earnestly appeals to the generous Irish people to give

the other £4,000 required

to maintain the Life-boats provided for the Irish Life-boat Crews in the service of humanity.

GEORGE F. SHEE, Secretary, Royal National Life-boat Institution,
Life-boat House, 42, Grosvenor Gardens,
London, S.W.1

Also standing by was the Royal Navy destroyer HMS *Tenedos*, and from about 3.30pm, when the gale had eased a little, until darkness two hours later, attempts were made to establish a tow between the lightship and the destroyer. Even when the lifeboat was twice able to pass the line aboard the casualty, the line parted. All the time, the three vessels were being swept by heavy seas.

Towing attempts became impossible once darkness had fallen and, with HMS *Tenedos* prepared to stand by all night, Coxswain Sliney decided to make for Cobh. He needed more rope and, more importantly, his wet and exhausted crew needed food. Harbour was reached at 9.30pm.

Robert Mahony, the honorary secretary, had been spending the time his lifeboat was away desperately trying to keep track of her progress. Fallen trees across roads and telephone lines down made his task virtually impossible. Finally, at 11pm on the Tuesday night he made contact with Cobh by telephone:

> I spoke to the coxswain. He told me the position, and I went back at once
> to Ballycotton and set out for Queenstown (Cobh) with a spare drogue,
> tripping line and veering lines, and changes of underclothing for the crew. It
> was twenty-three miles to Queenstown, and again a very difficult journey by
> night, dodging fallen trees. I arrived at Queenstown at three in the morning of
> Wednesday 12th, handed over the stores and returned to Ballycotton.

Although there were always three crewmembers tending the lifeboat while she was alongside in Cobh, some managed to snatch a little sleep while they were ashore. Early in the morning of the 12th Patrick Sliney and his men were once more negotiating heavy seas as they made their way back out to the lightship. The wind had dropped to some extent and fog had set in, but the sea did not seem to go down. This would be a day of standing by. HMS *Tenedos* had to take her leave, but the *Isolda,* an Irish Lights vessel, was on her way from Dublin.

The weather was in no mood to give respite. By the evening the forecast was predicting more gales. The lifeboat could not leave the lightship unattended overnight so she stayed, her crew out of food, utterly exhausted and constantly fighting a battle with gravity as their boat rolled and climbed and plummeted in the heaving Atlantic.

By daybreak, just after 7am, the coxswain was forced to return to Cobh as his petrol was getting low. By the time he and his crew reached harbour, they

An artist's impression of the long-awaited moment of rescue for the crew of the Daunt Rock lightship as the Ballycotton lifeboat makes one of several daring approaches alongside.

had been standing by for more than 25 hours. They were starving and soaked to the skin by the seas, which had been breaking over them throughout their vigil. There followed an anxious wait for fuel to be brought and it was not until late afternoon on the 13th that the lifeboat was back in the vicinity of the lightship. The weather had got noticeably worse and it was getting dark, but at least the *Isolda* had now arrived.

Her captain told the coxswain that he would not attempt a tow till the morning and was going to stand by all night. Patrick Sliney knew that, for the second night running, he would have to do the same. The south-easterly gale was growing in strength by the minute. At 8pm a giant wave burst over the lightship and swept away one of the two red lights hoisted at the bow to warn shipping that she was out of position.

With growing anxiety, the coxswain took the lifeboat round the lightship's stern where, with the help of the searchlight, he could make out the crew huddled in the stern, their lifejackets on and drenched by every sea as it broke over the vessel. Worse still, the wind was veering more southerly and, if it shifted any more to the west, the lightship could not avoid being swept on to the Daunt Rock itself, which was now only some 60 yards away.

Another brief exchange between the *Isolda*'s skipper and the coxswain determined that there was nothing that the support ship could do in such heavy seas, but that the men on board the lightship were now in grave danger and needed desperately to be taken off. It cannot have been a more daunting prospect for a lifeboat crew. The vessel was often hidden by breaking seas; when the searchlight did pick her out, she could be seen straining violently at her cable, plunging and rolling to impossible angles and thrusting her stern high into the sky. If that was not enough to ward off any approach, she was also fitted with anti-rolling chocks, which stuck out more than 2ft on either side of her, threshing the water menacingly as she rolled.

The least dangerous approach for a lifeboat would normally be to let go an anchor and to drop down with the wind towards the casualty. Patrick Sliney knew this was out of the question here because of the lightship's cable. Instead, having shouted instructions to her crew, he began a quick run in from astern of her and as he drew close to her port side, called on the crew to jump for the lifeboat. They only had about a second to make their leap – the time in which it took the coxswain to go from full ahead to full astern.

To remain alongside any longer would have been suicide; the lifeboat could at any moment be crushed as the ship rolled or swept towards the cable at the bow and instantly capsized. On the first approach one man jumped; on the second, no one dared. On the third, five leapt aboard together but on the fourth the lightship swung violently towards the lifeboat and crashed down on her, smashing the guard rails and damaging the fendering and part of the deck. The man working the searchlight leapt out of the way only just in time and miraculously no one else was hurt. The lifeboat drew back and ran in for a fifth time but neither of the last two men on board would jump. Instead, they simply clung to the rails, apparently frozen with fear.

The Ballycotton lifeboat crew, including four members of the Sliney family, photographed soon after they had come ashore from their 76-hour mission to the Daunt Rock lightship in February 1936. Left to right: Michael Walsh; Mechanic Thomas Sliney (coxswain's brother); Second Coxswain John Lane Walsh; John Sliney (coxswain's brother); Coxswain Patrick Sliney; Thomas Walsh and William Sliney, the coxswain's son, holding a dog. (*RNLI*)

For the next approach, Sliney sent some of his crew forward, something he had resisted up to now for fear of seeing them swept overboard. As the lifeboat came close, they reached up and grabbed hold of the two men, prising them from their perch and tumbling them unceremoniously on to the deck of the lifeboat. One of them cut his face badly as he hit the deck and the other injured his legs, but they were safe. The lifeboat crew did their best to tend their injuries as Patrick Sliney reported his mission accomplished to the *Isolda* and set a course for Cobh.

On the passage back, the strain of the last few days became too great for one of the lightship's crew and he became hysterical. It was all that the lifeboat crew could do to hold him down to prevent anyone being hurt or knocked overboard. The lifeboat eventually arrived at Cobh at 11pm. The crew spent the night there before setting off for Ballycotton the next morning, arriving home just after midday.

The Daunt Rock lightship herself did survive the storm. Her cable held long enough to allow the *Isolda* to attach a tow in more moderate weather on Friday 14 February. She arrived in Cobh harbour that evening.

The June 1936 issue of the RNLI's journal *The Lifeboat* deemed the Daunt Rock lightship rescue to be 'one of the most exhausting and courageous in the history of the lifeboat service'. The Ballycotton crew had been away from their station for 76½ hours, 49 of which had been spent at sea in bitterly cold rain and sleet and under constant onslaught from heavy seas breaking over their open lifeboat. Every man came back suffering from colds and salt-water burns and the coxswain from a poisoned arm. In the 63 hours between leaving Ballycotton and landing the survivors in Cobh, they had only had three hours' sleep.

Coxswain Patrick Sliney was duly presented with the RNLI Gold Medal for this service while his second coxswain, John Lane Walsh, and his brother and mechanic, Thomas Sliney, received Silver Medals. The remainder of the crew, consisting of his other brother John, his son William, Michael Walsh and Thomas Walsh, were given Bronze Medals.

Today, more than 70 years on from this epic rescue, the pride of the people of Ballycotton for what that 1936 crew achieved and for the medals they were awarded is still plain to see. The station itself exhibits many mementoes of the event, including a plaque on the exterior wall. It is not just history that continues to inspire the lifeboat operation in Ballycotton, the Sliney and Walsh families are still very much involved. Patrick Sliney's grandson Colin has recently retired from the crew, but his son Tom is now an active member and represents the fourth generation of his family to go out in the lifeboat. Two other descendants of the Daunt Rock rescue crew are also to be found aboard today's lifeboat, namely Michael Walsh, the mechanic, grandson of his namesake, and Redmond Lane Walsh, grandson of the second coxswain.

A proud wife and mother: Mrs Sliney accompanies Patrick and William Sliney to London for their Gold and Bronze Medal presentation. (*RNLI*)

RNLI Gold Medal winner Henry Blogg, coxswain of the Cromer lifeboat, worked as a fisherman. He rarely left Cromer, except on one occasion to go to Buckingham Palace to be presented with the Empire Gallantry Medal from King George V. (*Popperfoto*)

9

CROMER, 6 AUGUST 1941

In one of Henry Blogg's most famous Gold Medal rescues, six merchant ships from a wartime convoy run aground on Haisborough Sands and begin to break up. Two lifeboats from Cromer and the Great Yarmouth and Gorleston lifeboat rescue all the surviving crews from the six steamers in turbulent and very shallow seas. Of the 119 men saved, Henry Blogg rescues 88.

Henry Blogg was a man of magnificent deeds but very few words. His unequalled record of lifesaving and bravery at sea led him to become as famous in his day as any politician, sportsman or film star; yet he must have been a great disappointment to journalists trying to colour their accounts of his rescues with words from the great man. He felt, like many lifeboatmen before and after him, that whatever he told a newspaper would look boastful on the page, so it was better to say nothing.

It is rare for someone to gain such celebrity without ever living more than a few hundred yards from the house in which they were born, but Henry Blogg seldom left Cromer, except when invited to London to receive one of his many awards. On one occasion, during the centenary of the RNLI in 1924, he was summoned to Buckingham Palace along with six other RNLI Gold Medal holders to receive the Empire Gallantry Medal from King George V. When he returned to Cromer, his wife Ann was eager to know all about it, but all he would say to her was that he felt glad to be home.

Henry Blogg was born in 1876 to the unmarried Ellen Blogg who, with her other child, Mary, lived with her parents in a fisherman's cottage in Tucker Street.

He was sent to Goldsmith's School where he showed he was quick to learn and had an unusually retentive memory. He took no part in games, never learned to swim and was not good at defending himself against bullies. He left school at 11, by which time his mother had married John James Davies, who was second coxswain aboard the lifeboat. Henry worked for his stepfather, who had three fishing boats, and very quickly learned the skills required to handle an open sailing and rowing boat. In the summer he would supplement his income hiring out towels and bathing dresses at a penny a time and helping to get the horse-drawn bathing machines down to the water's edge in the cause of preserving the modesty of Victorian ladies.

By 1894, at the age of 18, Henry Blogg was deemed a strong enough oarsman to take his place as a member of the lifeboat crew. His stepfather, John Davies, retired as coxswain of the lifeboat in 1902 and Henry was elected second coxswain. By then he was a married man having wed Ann Brackenbury, a local girl, the year before. They had two children, a son who died in infancy and a daughter who tragically did not live beyond her early 20s.

Blogg bore his personal trials in the same way he handled the vagaries of the sea: with quiet steadfastness. He did not drink, did not smoke and did not swear, but even as a young second coxswain, he commanded the respect of his fellow crew. His modest humour and kindly nature endeared him to them; his skill as a seaman inspired their confidence; and his resolute decision-making, judgement and fearlessness made them look to him as a leader. In 1909 he was unanimously elected the new coxswain of Cromer lifeboat.

Even those who understood the qualities of Henry Blogg would never have predicted the extraordinary career he would lead during his 38 years as coxswain. They covered the two world wars, each of which caused east-coast lifeboats to attend an unprecedented number of emergencies. Blogg won his first Gold Medal rescuing survivors from a Swedish ship blown to pieces by a German mine in 1917. He went on to win two more Gold and four Silver Medals, a record of gallantry unlikely ever to be equalled. In all his years as a lifeboatman – he retired at the age of 71 – he had been involved in the saving of 873 lives. It was grimly appropriate that even at his death, seven years later in 1954, it should come after he had collapsed while helping three fishermen, two of them his nephews, whose boat had sunk in sight of Cromer Promenade.

A portrait of Henry
Blogg taken in 1942,
a year after he
rescued 88 men from
a stranded wartime
convoy. He was
coxswain of Cromer
lifeboat for 38 years
from 1909. (*RNLI*)

The Second World War was the busiest and the most dangerous period in Henry Blogg's career. When the lifeboat launched, he and his men were not only on the lookout for rogue waves, shoals and strengthening winds, there were mines, enemy planes and e-boats to avoid as well. Even if it was still the sailor's oldest enemy, the submerged sandbank – especially treacherous off the Norfolk coast – was often the ultimate cause of a shipwreck in those years, and wartime conditions such as coastal blackouts contributed greatly to these accidents. They also made search-and-rescue all the more challenging. Blogg won two of his Silver Medals rescuing crews from grounded merchant ships during the war, the first in 1939 when the *Mount Ida* was wrecked on the Ower Bank and the second when the *English Trader* ran on to Hammond Knoll in October 1941.

But it was in the summer of 1941 when the mettle not only of Cromer lifeboatmen but of the Great Yarmouth and Gorleston crew as well was put most sternly to the test. Before dawn on the morning of 6 August, a convoy of merchant ships, escorted by the destroyers HMS *Wolsey* and HMS *Vimiera*, was making its way south down the east coast of England, travelling through the narrow waterway which was kept buoyed, lighted and swept of mines and which, off the East Anglian coast, was known as E-boat Alley. This was where Allied ships came into the range of the fast German torpedo boats whose raids out of the captured Dutch ports had taken a considerable toll on merchant convoys.

The weather on this particular morning would have made life difficult for raiders. A gale was blowing from the north-west with a rough sea and squalls of wind and rain making visibility very poor. It would have been difficult in those conditions to make out the markers of E-boat Alley and as the convoy passed between the notorious Haisborough Sands to port and the Norfolk coast 8 miles to starboard, skippers could only follow the ship immediately ahead, trusting that the leading vessel was on course.

Somehow, though, they were steaming too far to the east and, at about four in the morning, the most easterly vessel struck the sands. Before she could give any warning, five more ships had ploughed into the bank, their stunned crews looking out over an expanse of boiling white water while their stationary vessels shuddered and groaned beneath them as wave after wave broke against the hull.

Whether the escorting destroyer captains thought at first that the vessels might refloat or that their own ships' boats would be able to rescue the crews without further assistance is unclear, but it was not until four hours after the groundings that lifeboat help was requested. By then, every available boat was required, the call having gone out to Cromer, Gorleston, Sheringham and Lowestoft. Cromer's No. 1 lifeboat, *H.F. Bailey*, launched from her boathouse at the end of the town pier soon after 8am. Henry Blogg and his crew had about 17 miles to cover and were first on the chaotic scene some 1 hour 40 minutes later. They had been told little other than that their presence was badly needed, so when they came across six steamers, the *Oxshott* and the *Deerwood* of London, the *Gallois* of Rouen, the *Taara* of Parmu, Estonia, the *Aberhill* of Methil, Fife, and the *Paddy Hendly* at varying stages of disintegration on the sands, they realised

Cromer's No. 1 lifeboat, *H.F. Bailey*, at the top of her slipway, built on to the end of the town's pier. (*Grahame Farr Archives*)

they had an enormous task on their hands. (*Author's note*: When the RNLI later checked the names of the ships as noted by the lifeboat crews they could find no vessel named *Paddy Hendly* in Lloyd's Register. As the men rescued from this ship were put aboard a destroyer with other survivors, there was no way of finding them again to check its real name. It remains an intriguing mystery as to whether a misunderstanding occurred or whether the lifeboat crew were perhaps intentionally given the wrong name.)

Where should Henry Blogg and his crew begin? An RAF aircraft flew overhead, the two destroyers stood off in deeper water and a whaler which had been launched from one of them was attempting to take men off the wrecked ships in the steep breaking seas that rampaged over the sands. Although, against all odds, the whaler had got most of the crew of the *Taara* to safety aboard one of the destroyers, 12 men from other ships were already dead, drowned in their attempts to swim to the warship's boat.

The ship appearing to be in most desperate trouble was the *Oxshott*. Henry Blogg could only make out her funnel and upper deck as white water cascaded over her. He could also discern a dark and shifting silhouette just aft of the funnel which, as the lifeboat drew close, revealed itself to be a huddle of 16 men, the entire ship's company, roped together and clinging on to each other or any remaining hand-hold on the engine-room roof. They could feel and hear their vessel breaking up beneath them and knew there was little time left. Blogg very soon realised that there was nowhere left on the ship's deck to secure a line from the lifeboat and that he had to get as close as possible to get the men off. A gaping crack had opened up in the iron plates on one side of the engine room and without hesitation the coxswain pointed his bow straight at it.

The artist Mick Bensley's impression of Henry Blogg wedging the *H.F. Bailey*'s bow in a crack in the *Oxshott*'s side to take the crew off the disintegrating ship.

More than any man, Blogg knew the strength of a lifeboat hull. This was not the first time that he would take the calculated risk of driving his boat right over the deck of a sinking vessel. This time, though, his plan was to wedge his bow into the crack in the ship's side so that he could hold the lifeboat there with the pressure of the engines while the survivors came aboard. Wave after wave fought to dislodge the lifeboat from her precarious position, but each time the coxswain drove her back and held her there while the 16 men scrambled aboard. All the while, seas would wash over the lifeboat and then abandon her so that she would crash down on the deck beneath her keel. It was obvious to the crew that she had been fairly badly damaged, although the extent of it they could not tell.

At last all the men were aboard and Blogg shouted to his mechanic to go hard astern. The lifeboat slid off the steamer's deck and back into the fury of the fast-ebbing shallows. With a grapnel line still attached to the *Oxshott*, the lifeboat was able to veer down on to the next ship aground, the *Gallois*. Her decks were still above water and Blogg was able to hold the lifeboat alongside, her head to the seas, while 31 men shinned down ropes or jumped aboard. One of them mistimed his jump and fell into the sea, but the Cromer crew were swift to pull him aboard, unhurt.

With 47 shivering survivors taking up every space aboard the lifeboat and weighing her down in the shallow water, there was an urgent need to offload them. Blogg headed off the sands and steered towards one of the destroyers where the men were exchanged for a few tots of heartening Royal Navy rum. As the lifeboat pulled away from the destroyer's side, her crew caught sight of the familiar shape of Cromer's second and smaller beach-launched lifeboat, the *Harriot Dixon*, which had just arrived on scene and was heading for her sister boat to find out what was required of her.

When the two lifeboats were alongside each other, Henry Blogg asked his nephew, Second Coxswain Jack Davies, to go aboard the *Harriot Dixon* and take command. His experience that morning of taking two ships' crews off the sands would be invaluable in the further rescue attempts. Blogg, aboard the *H.F. Bailey*, now turned towards the next ship, the *Deerwood*, whose bridge was the only part of her that was visible above the crashing waves and the only refuge for her 19-man crew. For the second time that day he drove over the bulwarks of the submerged vessel and held his position against the bridge by working the

Cromer's second lifeboat, *Harriot Dixon*, and her crew at sea in 1942.

engines, all the time risking more damage to the lifeboat's hull. With all 19 men aboard, he pulled away and looked about him to see where he should go next.

He headed for the *Aberhill*, lying in the surf with her back broken. Then he noticed another lifeboat along the steamer's lee side, amidships. She was the *Louise Stephens*, having battled her way up from Great Yarmouth and Gorleston and her coxswain, Charles Johnson, and his crew had got lines aboard the ship to secure her long enough for the entire crew of 23 to be taken off. Meanwhile, Cromer's No. 2 lifeboat had made for the *Taara* where eight men were still waiting to be rescued. This ship had also broken her back and both her stern and her bow were under water. Using a similar technique to Henry Blogg's, Jack Davies held his boat against the steamer's bridge, working the engines and keeping his head to wind and sea while the men clambered aboard.

The tide was still ebbing, but there was still a sixth ship, the *Paddy Hendly*, to attend to. It was Henry Blogg's turn again, and once more he steered the lifeboat alongside and held her there while the last 22 men leapt aboard. Now the water had become so shallow and the lifeboat so weighed down with survivors that she began to bump the bottom as she pulled away from the wreck. Then the lifeboat came to a sudden halt as the entire length of her keel dug into the sand. Any sea breaking over her now would capsize her and wash every man aboard away. Coxswain Blogg watched helplessly as a very large wave loomed up on the port side. To everyone's huge relief, it broke just before it reached them and lifted the lifeboat off the bottom. The engines were thrust full ahead and in 20 yards they were back in the deep water.

HRH the Duchess of Kent, Princess Marina, President of the RNLI, on a visit to Cromer on 23 May 1945, where she met Henry Blogg and his crew.

Harriot Dixon is launched from her boathouse on Cromer seafront, 1935.

It was 1 o'clock in the afternoon. Three lifeboats had saved the lives of 119 men. Two more lifeboats, from Sheringham and Lowestoft, arrived on the scene to find that the rescue was over and so turned for the long and gruelling journey home. *Harriot Dixon*, the Cromer No. 2 lifeboat, transferred her survivors to a destroyer and headed back to Cromer beach. The larger Cromer lifeboat could not have been rehoused up her slipway in that weather, so she followed the Great Yarmouth and Gorleston lifeboat back to Gorleston Harbour. Before she got there she met a destroyer and put the 41 men from the *Deerwood* and *Paddy Hendly* aboard her.

By the time the *H.F. Bailey* tied up alongside at Gorleston, she had been at sea for nine hours. Blogg and his crew made their way to the Mariners' Home for food and hot baths, possibly reflecting on the extraordinary seaworthiness of their boat in spite of the damage that day's work had inflicted on her. She had three holes in her bow; an 8ft portion of her stem had broken away along with a 20ft length of fender. The brass bolts that had held the stem in place had been driven inwards through 8in of oak and had pierced the air-cases inside the hull.

This epic rescue not only yielded Henry Blogg's third Gold Medal, but also Silver Medals for Jack Davies (who took command of Cromer No. 2 lifeboat on the sands), and for the Great Yarmouth and Gorleston coxswain, Charles Johnson. Leslie Harrison, coxswain of Cromer No. 2 boat, his mechanic Harold Linder, Blogg's mechanic Henry Davies, and the Great Yarmouth and Gorleston mechanic George Mobbs all received the Bronze Medal.

Henry Blogg was further honoured for this rescue when he received the British Empire Medal, the news of which came only very shortly after he had heard that the Empire Gallantry Medal he had been awarded in 1924 was to be replaced with the George Cross. He remains, to this day, the only lifeboatman to have received this award.

10

HUMBER, 6 JANUARY 1943

Robert Cross, legendary coxswain of the Humber lifeboat, based at Spurn Point, spends a gruelling night in an easterly gale among the wartime defences and sandbanks at the mouth of the Humber. He first rescues 5 men adrift on a floating gunnery platform, then he takes 19 men off a Royal Navy trawler which had run on to a shoal.

U nless you are a seafarer or a hardened wildlife enthusiast, you might be hard put to locate Spurn Point on a map of England. It is certainly not a place you would happen upon by accident. It consists of a 3½-mile long spindle of sand and shingle that curls south and westward into the mouth of the River Humber at the southern end of a sweep of the Yorkshire coast which begins at Flamborough Head.

The narrow spit is only 55 yards wide at certain points and in recent years the fragile roadway along its length has been washed away by high tides and heavy seas on a number of occasions, isolating its more bulbous head from the mainland. Any visitor to this beautiful but desolate spot may not think that too important until they arrive at the very tip where they will find evidence of more human activity than they probably expected.

Spurn Point has always provided a highly strategic base for Humber pilots who are able to meet vessels entering the estuary to guide them into Grimsby, Immingham, Hull or Goole. Their operation today is part of the Associated British Ports' establishment maintained at Spurn to control all the shipping entering and leaving these ports. Until the need for a lighthouse was supplanted by electronic navigation equipment in 1986, Trinity House had always had a major part to play and was originally responsible for the pilots and maintaining the light. It also ran a lifeboat station at Spurn from 1810 until 1911 when the responsibility was handed over to the RNLI.

From the outset, the RNLI realised that the only way to guarantee a lifeboat's availability 24 hours a day, 7 days a week, in this remotest of places was to keep a crew permanently housed at Spurn. Unlike every other station of the RNLI, the

How Spurn Head appeared at the time of the 1943 rescue, showing housing provided by the RNLI to allow a lifeboat crew to be constantly available for Humbermouth incidents. (*RNLI*)

crew would have to be full-time as there was no other way of earning a living on the peninsula. Houses were built for them and their families and a unique lifeboat community was established which endures to this day.

One man, Robert Cross, knew the station both before and after its adoption by the RNLI. He joined the crew at Spurn in 1902 and stayed there for six years until he had saved up enough money to buy a share in a drifter when the herring industry was at its height. Barely a year into his new occupation, however, a tragic event affected him profoundly and changed his life forever. Several cobles had been caught out in a gale and he volunteered to go out with the Flamborough lifeboat to look for them. Two of the cobles were lost, one of them with Robert Cross's brother and two nephews aboard. From that day he decided his mission in life was to save people from the sea and turned to the RNLI for a permanent post. This was how he came to be appointed the first RNLI coxswain at Spurn Point, a position he would hold until November 1943 and which would make him one of the most decorated lifeboatmen to have lived.

His career as coxswain, similar to that of his famous east-coast contemporary Henry Blogg, spanned both world wars and was exceptionally busy, not just because of war casualties but because of the immense amount of traffic using the waters around the Humber Estuary in times of war and peace. By the declaration of the Second World War, Robert Cross's record was already an extraordinary one with two Silver Medals and a Bronze to his name, all for rescues to vessels trapped on the treacherous sandbanks that litter that part of the coast. In October 1939 he again risked life and limb saving seven men by breeches buoy from their trawler, aground on the Inner Binks, a notorious shoal close to Spurn Head. He had not yet been presented with the third Silver Medal this rescue had earned him when, the following February, he and his crew were called to another trawler, the *Gurth*, foundering on a shoal to the south of the estuary. In gale-force winds, snow and total darkness, Robert Cross, in spite of a vicious ebb tide, repeatedly manoeuvred his lifeboat alongside the swamped trawler and took off her nine-man crew. The RNLI awarded him the Gold Medal this time, and he also learned that he would become the first person ever to receive the newly instituted George Medal.

A year later, in February 1941, Cross was again involved in a medal rescue, this time a bar to his Bronze, when he took the lifeboat into an area thick with enemy mines to rescue the crew of a ship flying barrage balloons, which had run

Coxswain Robert Cross, holder of two Gold, three Silver and two Bronze RNLI bravery medals in addition to the George Medal. (*T.C. Turner*)

aground on the southern shore of the estuary. By now the mouth of the Humber had become a navigational nightmare. Hidden sandbanks were the least of any skipper's concerns. German mines could be lurking anywhere, dropped by aircraft for whom the Humber was a prime target. Much had been done to guard the estuary from attacks from the air and by submarine, but these defences were as much a hazard as a help to the local lifeboat and all friendly shipping.

Two steel forts, built on piles above the sandbanks, overlooked the estuary from either side. Between them a defence-boom ran right across the river mouth, a 3-mile line of buoys with massive balks of timber chained between them, each one bristling with steel spikes. Suspended beneath these timbers were steel nets reaching to the sea-bed, ready to trap any submarine raiders intent on causing havoc upriver. All the Humber's legitimate traffic had to pass through a 100-yard gap in the boom, between the gate-ships, which could lower and lift the nets, and keep to a narrow channel, which would be cleared morning and evening by minesweepers.

Guns, searchlights and mines were hidden in the dunes of Spurn Point and iron hedgehogs were set on the sand below high-water to prevent enemy aircraft from landing on the beach. Anchored just inside the defence boom lay the Phillips defence units: huge iron buoys mounted with anti-aircraft guns together with military trawlers flying balloons.

However much as these defences may have deterred a human enemy, they only served as playthings for the Humber's much older adversary: the easterly gale. On the evening of 6 January 1943 a brutal, freezing wind carrying horizontal snow came whipping off the North Sea. For anyone daring to put their head out of doors at Spurn Point, the pummelling of the gale and the crashing of an angry sea meeting the strong ebb tide out in the estuary would be all that they could hear. Robert Cross could well imagine what the people out in the thick of it, manning the gate-ships and the defence units, were having to endure. Then, shortly before 8pm, he knew that he and his crew would have to go out there, too.

The No. 1 Phillips defence unit had broken adrift from its mooring and had run on to the north side of Trinity Sand, inside the boom. The lifeboat station at Spurn was positioned just on the seaward side of the boom so it meant that Robert Cross would have to take his twin-engine, 45ft 6in Watson class lifeboat, *City of Bradford II*, through the gap to reach the stranded gunners. As they headed towards it, the crew of one of the gate-ships attracted their attention and, using a megaphone, was able to tell them they could turn again for home; a tug had got to the unit and towed it to safety. The lifeboat crew were thankful for such a short mission on a night like that; they had been at sea for less than half an hour.

Just as the laborious process of rehousing the lifeboat at the top of the slipway was complete, the telephone rang again. A trawler was aground on the Binks, on the south-eastern side of the head. Robert Cross knew, with an ebbing tide, that the lifeboat would not be able to get close to her and, cold and miserable though her crew would be, that the trawler would lie quiet on the sands until the flood tide. He sent his crew back home, advising them to get some sleep as they would be needed in the early hours of the morning when the tide came in.

The 45ft 6in Watson class lifeboat *City of Bradford II*, stationed at Spurn Point between 1929 and 1954, launches into the Humber. (*P.A. Vickery*)

The gale would not allow them to wait until then. At 10.50pm the telephone rang in the coxswain's house again. It was the port war signal station to say that another of their Phillips defence units, No. 3 this time, had broken adrift from her moorings and was entangled in the boom on its inner side. The crew were summoned from the beds they had just crawled into and hurried down to the lifeboat station. From there they could see red distress rockets shooting skywards from the defence unit and the men on board making desperate signals for help. At any moment the ebb tide could sweep them over the boom and out to sea.

Searchlight operators from the shore directed their beams on to the scene as the lifeboat approached from the seaward side. They were of arguable value to the coxswain as they tended to blind him and his crew as often as they showed the way. Robert Cross nevertheless succeeded in placing the bow of the lifeboat against the boom four times, enough to allow all five gunnery men to jump aboard. The yard-long spikes protruding from the boom had done the job they were put there for, however, inflicting serious damage to the lifeboat's stem and bow planking.

Fortunately, the damage was by no means enough to put an RNLI lifeboat out of action and Robert Cross headed back to the station to land his survivors. By now there was no point in rehousing the lifeboat as she would soon be needed for the stranded trawler, so he tied up alongside a patrol vessel to wait for the flood tide.

Just after 3am the next morning, the lifeboat cast off, rounded the head and, despite the heavy snow showers and pitch darkness, found the Admiralty-owned trawler *Almondine* lying on her side on the sands. A strong spring flood tide was swirling over the Binks at a rate of 6 knots and seas appeared to be breaking from all directions. An unequivocal signal flashed from the trawler's skipper that he wanted her crew taken off. His port side was under water and heavy seas were constantly enveloping the vessel.

The coxswain began his first approach, head to tide, aiming to come along the trawler's lee side. He managed this and his crew passed a line to the trawler where it was made fast. Then the tide caught the lifeboat and swung her sharply round; the rope broke, the lifeboat heeled over and her mast crashed against the *Almondine*, snapping the mast like a matchstick. With his radio now out of action, Robert Cross prepared for his next run in. In all, he made 12 approaches, dashing in swiftly, first from the trawler's bow end, then from her stern. All the time waves were sweeping over the lifeboat, hurling her about and occasionally forcing her hard against the trawler. This only added to the earlier damage to her stem and planking, which was now holed above the waterline. Sometimes the lifeboat pulled away from the trawler empty-handed; at others one man was able to jump or be wrenched aboard; occasionally as many as three came on board at once.

After 45 minutes, 19 men had reached comparative safety aboard the lifeboat. Only the skipper and officers remained on the trawler, which was now showing signs of re-floating. She was more upright and was shifting with the seas. Her skipper hailed the lifeboat and asked the coxswain whether he thought they should come off or stay aboard. Before Robert Cross could answer, the trawler's lights went out and he lost sight of her in a blizzard of driving snow.

For the next hour and a half, the lifeboat, with the help of her searchlight, scoured the sandbanks and entrance to the Humber for the vanished trawler. They found nothing and eventually returned to the station to land the 19 bedraggled survivors. Robert Cross immediately telephoned the port war signal station and learned to his relief that a tug had just reported finding the *Almondine* drifting in the entrance to the Humber and had taken her in tow. Her remaining crew were all accounted for.

The award of a bar to his Gold Medal was the least the RNLI could do in recognition of Robert Cross's courage and 12 hours of exertion that night. His reserve motor mechanic, George Richards, received the Silver Medal and the five other crewmembers the Bronze. Robert Cross decided to retire later that year; he was 67 years old and as well as the seven RNLI gallantry medals to his name, he had played a part in saving the lives of 453 people.

11

MOELFRE AND HOLYHEAD, ANGLESEY, 2 DECEMBER 1966

Richard Evans, coxswain of Moelfre lifeboat, and Harold Harvey, Inspector of Lifeboats for the North-west, win Gold Medals off the north coast of Anglesey, rescuing 15 men from the freighter *Nafsiporos* in hurricane-force winds.

With the useful benefit of hindsight, some might have concluded that the drama of rescuing 15 men from the crippled Greek freighter *Nafsiporos* in a hurricane barely a mile off the North Wales coast need never have been acted out. After all, the 1,287-ton ship, with the four men who refused to abandon her, ended up moored safely alongside the Liverpool dock at the end of the momentous episode.

As it was, three lifeboats, one from the Isle of Man and two from Anglesey, would encounter some of the worst weather ever experienced in the history of the RNLI and both the Welsh boats would come perilously close to disaster along the ship's side for the sake of her crew. For the men at the helm of both these lifeboats, life afterwards could never be quite the same again; Lt Cdr Harold Harvey, the Inspector of Lifeboats for the North-west, who happened to be visiting Holyhead when the lifeboat was called out, became the first RNLI inspector to win a Gold Medal for bravery. Richard (Dick) Evans, coxswain of the Moelfre lifeboat, gaining his second Gold Medal, had instantly earned the reputation of 'RNLI legend', to sit alongside names such as Robert Cross of The Humber and Henry Blogg of Cromer.

If Harold Harvey's participation in the events of 2 December 1966 had been coincidental there was, by contrast, something almost pre-ordained about Dick Evans's involvement that evening. Richard Evans, born in Moelfre in 1905, was destined, through family circumstances and an all-consuming passion for the sea, to be a mainstay of the local lifeboat crew. His father, a sea captain, crewed the lifeboat when he was at home on leave. Both his grandfathers had also been on the crew, one of them as second coxswain. Dick himself spent his childhood, when Chapel and a strict school regime would allow, as crew on his grandfather's 20ft sail-driven fishing boat, dreaming of the day that he would take command of his own ship.

That day came earlier than he could ever have imagined. He joined the crew of a coaster at the age of 14 and then moved rapidly up from able seaman, second mate, chief officer to skipper of the 320-ton *Colin* not long after his 23rd birthday. His first outing on the pulling and sailing lifeboat had come when he was only 16, when the crew were called to a schooner adrift in a gale and rescued the men aboard her.

An offer Dick Evans found he could not refuse, however reluctant he was to forfeit his life at sea, presented itself when his uncle, John Matthews, took up the job of full-time coxswain at Moelfre and asked Dick to run the family butchery business in his stead. It meant, at least, that he could spend more time on lifeboat duty and he progressed over the next 26 years from bowman to second coxswain and eventually to coxswain when his uncle retired in 1954.

Twelve years later, on 2 December 1966, Dick Evans was still coxswain of the lifeboat. It had been a morning at sea that none of his crew would forget in a hurry, not even Second Coxswain Murley Francis, Mechanic Evan Owens, Bowman Hugh Owen or crewmember Hugh Jones, who had been with him on the day in 1959 when he won the Gold Medal rescuing the crew of the *Hindlea* in 40ft waves and 90mph winds. Dick was now 61 years old and the strain of being at the helm of the RNLB *Watkin Williams* for six hours in a hurricane was understandably making itself felt. It had seemed that all hell had broken loose that morning as a cyclone had unleashed a vicious assault on the shipping off the busy north coast of Wales. Mechanic Evan Owens recalled how it all started:

The early hours of December 2 were wild and stormy. The wind from the North was gusting to hurricane force, with huge seas roaring in the Swnt at Moelfre, sending spray and foam up into the village. It was a night when lifeboatmen are uneasy, thinking of the storm and of the many things that can cause a distress call. As I listened to the roof slates rattling in the gale at Lifeboat House, the howling kept me awake thinking of those men out at sea. I had been one myself for many years.

At 6am Dick Evans the coxswain rang. He had received a message from Holyhead that the motor vessel, *Vinland*, had engine failure twenty miles north of Point Lynas and required assistance, a tug if possible, so would we

Coxswain Dick Evans at the helm of Moelfre lifeboat, *Watkin Williams*.

stand by to launch if a call should come? I opened my back door and closed it quickly, as the cast iron gutters came crashing down. The village was lit up by flashes, as the overhead electricity cables shorted. Everything movable was blowing around wildly and chimney pots and slates crashed down. Every experienced lifeboatman knows the feeling; one just knows it is going to happen.

At 7.30 the call came. The maroons exploded in a shower of green and white sparks, adding to the eerie scene. The men rushing to the lifeboat had to crawl along the wire fence, such was the force of the wind. They pulled on oilskins and lifejackets, the engines roared into life, the doors of the boathouse rolled open, the securing hooks were knocked open and with propellers spinning the *Watkin Williams* slid down the slipway, slamming in to the sea at nearly thirty miles an hour and causing a huge bow wave to cascade along the deck. The mast-exhaust was raised, the radio aerial raised and then at full speed the lifeboat faced the huge seas coming roaring through the Swnt and past Moelfre Island, rising and falling as she mounted each crest of breaking foam, then sliding down into the uncanny calm in the trough between walls.

It had taken them two-and-a-half hours to steam 7 miles into the north-westerly Force 10 towards the *Vinland*'s position. Then they had altered course towards another motor vessel, the *Grit*, in even greater danger with broken steering gear and only 3½ miles from a lee shore. That was another three hours' steaming. Richard Evans stood by her while feverish efforts were made to rig a temporary repair. The ship radioed eventually that she was once again under control. Meanwhile, the coxswain learned that another ship, *Pacific North*, was standing by the *Vinland* and a tug was on the way to her. He was able to turn for home.

Now the crew were thankfully back at the station. The lifeboat was halfway up the slip and the men were just beginning to realise how much they needed warmth and dry clothing and, above all, food after their gruelling morning at the mercy of the elements. They became vaguely aware of a telephone ringing in the boathouse above them. The winchman hurried down the slipway and said something to the coxswain. The phone call was for him; it was the Coastguard at Holyhead. When he got back to the boat with the news that they were wanted at sea again, this time round by the Skerries at the north-west tip of Anglesey, there was utter disbelief among the crew.

David Evans, the coxswain's son and a regular crewmember, recently remembered that moment: '*Holyhead*? We could not believe they wanted us to go all the way to Holyhead after it had just taken us three hours to cover just 3½ miles.' But nobody argues. The boat is refuelled and soon after 2pm, she buries her bow once more into the turbulent grey and white water at the foot of the slipway.

The 28-year-old skipper of the *Nafsiporos*, Angelo Katsovufis, had been faced with a difficult decision, barely a month into his first command. He was on passage to Belfast, having offloaded his cargo in Liverpool, when the weather worsened and his ship, with little ballast in her hold, began to pitch and roll to an alarming degree. Should he turn back and lose valuable passage time, continue on his voyage or seek the nearest shelter and ride out the storm at anchor?

The Isle of Man lay ahead of him so, choosing the last of these options, he ordered course to be set for Ramsey. With considerable difficulty the ship was guided through the worsening storm towards the comparative shelter of the Manx coast. The *Nafsiporos* was so light that every time her bow plunged into a trough, her propellers and rudder rose clean out of the water and the helmsman lost all control.

For a night the Greek ship lay at anchor, but her crew were far too anxious to sleep. Their vessel was behaving like a wild steer, tethered against its will, bucking and twisting with increasing vigour until at daybreak on 2 December it was clear that the anchor was dragging and that the ship was being swept by the north-westerly storm out into the Irish Sea. Her engines, in use in an attempt to hold her position, screamed ineffectually every time the propellers lifted from the sea. This caused them to overheat and before long they ceased to work.

Even before the power went, the *Nafsiporos* radio operator had signalled the first Mayday. By 8.20am, Douglas lifeboat *R.A. Colby Cubbin No. 1* had launched from her slipway and was making her way across a moving mountain range of sea towards a position the Greek ship had given as 12 miles south of Douglas Bay. After nearly three hours of steaming the lifeboat received news that an RAF Shackleton plane had the *Nafsiporos* in view 25 miles from Douglas Head. Not long afterwards the coxswain, Robert Lee, caught sight of the aircraft and altered course towards it. He never saw the Shackleton again and, search as they would for several hours more, they never found the *Nafsiporos*. Eventually news came over the radio that Holyhead lifeboat was in contact with the freighter and Coxswain Lee knew he was no longer required.

There were nine men on board the 52ft Barnett class lifeboat RNLB *St Cybi (Civil Service No. 9)* when she hurtled down the Holyhead slipway at 10.30 that morning. It is difficult to know quite how her coxswain, Tom Alcock, felt

Holyhead's 52ft Barnett class lifeboat, *St Cybi (Civil Service No. 9)*, on her slipway. William Jones, veteran of the *Nafsiporos* rescue of 1966, later became coxswain of the lifeboat and won the RNLI's Silver Medal when he and his crew saved the four people aboard the yacht *Pastime*, floundering in mountainous seas 30 miles south-south-west of the Skerries lighthouse. (*Liverpool Daily Post*)

as he steered for the harbour entrance towards the jagged white-flecked horizon of a boiling Irish Sea. He had only recently taken up his position as full-time coxswain at the station. Although a Welshman, he was from Rhyl, thus by no means a local, and had been promoted to coxswain by the RNLI after serving as bowman, first in his home town, then with the only full-time crew in the country at The Humber. Here was a shout that could make or break his reputation with the crew. As if further scrutiny were needed, by pure accident he also had on board the lifeboat inspector, the man responsible for his appointment, who had arrived at the boathouse minutes before the launch. What could he say but 'yes' when he asked if he could come too?

Although he had the highly experienced Mechanic Eric Jones and Second Coxswain Will Jones aboard, it had not been possible to summon all the regular crew. The storm had brought down all the telephone lines and the maroons had been inaudible to most above the wind. One of the oilskin-clad figures beside him was an 18-year-old boy, Graham Drinkwater, who had never been out on the lifeboat before (he would himself one day become coxswain of Holyhead lifeboat).

It took the Holyhead crew three hours in appalling conditions to locate the *Nafsiporos*. Eventually the lifeboat was spotted by a Shackleton, which was able to guide her in the direction of the casualty. The sheer height of the waves, estimated at 35ft between crest and trough, would have made the chances of finding the ship without the help of the aircraft very remote. For the first hour or so of her attendance, the lifeboat crew were witness to a daring attempt by a Russian timber ship, the *Kungurles*, to tow the *Nafsiporos* to safety. At considerable risk to the crews of both ships a line was attached when they were about a mile north of the Ethel Rock buoy, but in the struggle to bring the Greek ship's head to sea, a wave struck her and the heavy wire towing cable parted.

Now the two ships were too close to land to try to reconnect the tow. *Nafsiporos* let go her port anchor, but it failed to hold. She continued to drag until she fetched up less than a quarter of a mile to the west of West Mouse Rock in 6 fathoms of water with the Anglesey coast barely half a mile to leeward. At this point an RAF helicopter from Valley had made it to the scene and was attempting to hover over the ship. But with the vessel rolling 35 degrees either side of the vertical and the wind unabated at 100mph, gusting to 120mph, a winchman would not have had a chance of getting safely down to her deck.

As far as Coxswain Alcock could see, it was a lifeboat or nothing which would save the freighter's men. It was at about 4pm, with the light failing fast, that he made his first approach. He steered around the ship's stern in an attempt to come in along the starboard side where a ladder had been placed. Unfortunately, only a few feet forward of the ladder, the ship's lifeboat, which had earlier been turned out on her davit, was now hanging vertically, held only by the stern fall and swinging wildly as if to ward off any approach. Will Jones, the second coxswain, positioned in the bow, described his experience:

For a moment I thought we'd had it. I thought the *Nafsiporos* was going to roll over us. The sea had lifted her high in the air and her screw was churning round above our heads. I really thought she was going to crash down on us. It was frightening. A wave slammed us against the steel plates of the *Nafsiporos* and we had to sheer away. As we circled, all of us were looking for a safe approach but that ship's lifeboat was always in the way. We shouted through the loud-hailer for the crew to cut it away, but either they couldn't hear or couldn't understand our Welsh accents.

Watching this first approach with keen interest was the crew of the Moelfre lifeboat. It had been just before sunset in a fleeting moment of sunlight that Dick Evans and his exhausted men had first glimpsed the *Nafsiporos* after a murderous passage at full speed from their station. Dick Evans later gave a memorable account of that voyage:

I knew the rocks, the set and drift of tides, the currents. But that day the sea was like a foreign country. With the leaping and plunging of the lifeboat, the compass was swinging wildly. I could see nothing. The sea was being blown into clouds of spray and visibility was nil. We had to run on dead reckoning.

The waves were like nothing I'd ever been told about. We climbed perpendicularly and we went down the same way. I was afraid every wave was going to send us somersaulting on our back. There would have been no hope for any of us then, we would have disappeared forever.

As the lifeboat ploughed on, Dick attempted to give some warmth and encouragement to his crew by ordering the lifeboat's rum supply to be cracked

A painting by Michael Turner showing Holyhead lifeboat alongside the *Nafsiporos* at the moment when the ship's lifeboat crashed down on to the foredeck in 1966. (*Michael Turner*)

Three of the main players in the *Nafsiporos* rescue: from left, Dick Evans, Harold Harvey and Tom Alcock.

open and passed around. Sadly, neither bottle made it to any of the men's lips. Both were smashed as the second coxswain was thrown around the cabin while he attempted to bring them up on deck. The smell of spilt rum that permeated the boat only added to the nausea nagging away at each man's empty stomach.

Far more serious damage occurred 8 miles west of Point Lynas. The coxswain realised that the lifeboat had begun to bury her bow far deeper into the waves than was normal. Then he saw what looked like two gaping holes in the foredeck. Immediately he reduced speed and asked his second coxswain, Murley Francis, and his son to go forward to inspect the damage:

David Evans, who
accompanied his father Dick
on the night of the *Nafsiporos*
rescue, outside the house
where his parents lived and
where a plaque in memory
of Dick Evans has been fixed
over the door. (*Edward Wake-
Walker*)

Two retired coxswains of the
Holyhead lifeboat: Will Jones
(right), who was second
coxswain aboard the RNLB
St Cybi (Civil Service No. 9)
for the *Nafsiporos* rescue and
Graham Drinkwater (left),
who, at 18, was out that day
on his first 'shout'. (*Edward
Wake-Walker*)

It was dangerous; they could easily have been swept overboard. But, if I sent anybody, I had to send my own son. If I had sent someone else and they would have been washed overboard I'd have to live with the reputation of saving my own son at the expense of my crew. It had to be David. Murley I could not have kept back anyway. I watched with my heart in my mouth while they crawled forward and I don't think I breathed again until they were back.

Their news was that the force of the sea had ripped away the two deck ventilators and the bow of the lifeboat was full of water. They had managed to plug the holes with spare sou'westers and the mechanic's oilskin trousers. It was not much later when the *Nafsiporos* hove into view.

One of the two lifeboats on the scene had damaged her radio, so Dick Evans did not know whether the attempt they had just witnessed by the Holyhead crew to take men off had been successful or not. He decided, however, that it was now his turn to make an approach. His son, David, remembers to this day the extraordinary view right along the underside of the *Nafsiporos*'s keel as her stern was thrown high above the lifeboat. Only recently has he confessed to pleading with his father, albeit under his breath, not to attempt an approach in such conditions.

With a 5-knot ebb tide running east to west, the anchored ship would not keep head to sea and exceptional seamanship was needed to keep engines and rudder in the right position to prevent the current from pushing the lifeboat away from the vessel's side. Ignoring his son's misgivings, Dick Evans drove the lifeboat on, the first time making contact with the vessel but needing to sheer away quickly to avoid the swinging ship's lifeboat. He came round again and once more manoeuvred alongside, but the terrified Greek crew, mesmerised by the lifeboat, which was one moment lifting high above the deck and the next plunging into a 20ft abyss below, would come nowhere near the ladder. Dick Evans eventually pulled away from the ship's side to avoid unnecessary damage.

Back aboard Holyhead lifeboat, after his first jarring attempt to get alongside, Coxswain Tom Alcock had made an unusual, if pragmatic, choice. He had handed the helm to the lifeboat inspector beside him in the wheelhouse. Second Coxswain Will Jones, up in the bow ready to grab survivors, would probably have looked surprised to see his coxswain making his way forward to join him, while through the windshield he could see the shape of Harold Harvey standing at the helm. Alcock perhaps believed, with his regular bowman off sick, that his own experience as bowman from his previous job with the RNLI was what was wanted now.

At about 4.45pm, more than half an hour after the first attempt by Holyhead lifeboat to get alongside, Harold Harvey, constantly adjusting engines and rudder, coaxed the lifeboat towards the jumping ladder. It looked as if the *Nafsiporos*'s crew had finally realised that they had only one likely means of survival; a man could be seen outboard, clinging nervously to the top of the ladder. With crucial timing, the two men in the bow of the lifeboat grabbed hold of him and, using all of their combined strength, they just succeeded in dislodging him from the ladder as he clung like a limpet to his perch.

In spite of the wild movement of the two vessels, four more men were dragged aboard the lifeboat in similar fashion as Harold Harvey fought to hold his position. It was impossible to prevent the lifeboat's bow from occasionally striking the dangling ship's boat and suddenly, just as the fifth man had been brought aboard, the inspector saw the rope from which the boat was hanging give way. As he put the engines hard astern, he yelled a warning to the men on deck. They threw themselves aft as the boat came crashing down on the forward end of the lifeboat. It landed upside-down, spilling all its gear, including the oars which protruded through the open wheelhouse windows. The lifeboat's mast carrying the exhaust was destroyed along with the forward stanchions of the guard chains. Miraculously, not one of the 14 crew and survivors aboard the lifeboat was seriously injured. As the lifeboat backed away from the freighter's side, her crew watched incredulously as the additional cargo of an upturned boat teetered over their port side. To everyone's relief, the momentum of the lifeboat and the intervention of a wave breaking over the bow helped to dump the boat over the side.

There was now a considerable amount of debris in the water beside the *Nafsiporos*, but that did not deter Dick Evans from taking his turn at getting alongside. Most of his crew were now lining the port rail, ready to grab survivors as they clambered on to the ladder. The Moelfre crew seemed to ignore the huge risk they faced of being crushed as the two vessels came forcefully together time and again. Dick Evans later described the experience as being 'like a ball thrown against the gable end of a house'. Somehow he held the lifeboat there long enough for ten men to come aboard.

Coxswain Dick Evans with the crew of the *Nafsiporos* after the rescue. (*Liverpool Daily Post and Echo*)

The last four were not going to leave, however; Dick recalls: 'I was shouting my head off, probably in Welsh, for them to come to the lifeboat, and they were shouting back, in Greek, that they were not leaving the ship.'

The two lifeboats now set a course for Holyhead to land the men they had managed to take off the *Nafsiporos*. By the time they set an unsteady foot ashore, the Moelfre crew had not eaten for 23 hours. Dick Evans had been at the helm for an uninterrupted 12½ hours. The Holyhead crew gave themselves time only for a cup of tea before they set out again to stand by the *Nafsiporos*, which was still in grave danger. When the Dutch tug *Utrecht* arrived on scene that night the lifeboat, with the help of the RAF, was able to guide her into the confined waters and recommend an approach for attaching a towline. Thus the vessel and the four men aboard her were eventually saved.

As well as the two Gold Medals awarded to Dick Evans and Harold Harvey, the RNLI saw fit to present Silver Medals to Coxswain Thomas Alcock, Mechanics Eric Jones of Holyhead and Evan Owens of Moelfre and Bronze Medals to every other lifeboat crewmember who had taken part in the momentous rescue.

Dick Evans, BEM, retired as coxswain in 1970, but his second remarkable career with the RNLI was only just beginning. Most lifeboat coxswains would sooner face 40ft waves and a blizzard than stand up before an audience to talk about their adventures at sea. Dick Evans, however, was a natural orator who was able to hold an audience rapt as he recalled in lilting, almost poetic tones his absolute moments of truth alongside stricken casualties. He was invaluable to the RNLI in public relations and was responsible for many wills to be altered in favour of the lifeboat service. He was twice invited to speak at London's Guildhall, the second time at the RNLI's 150th anniversary dinner. There was not a dry eye in the house. Neither was there when Dick was ambushed by Eamonn Andrews to appear as the subject of the television programme *This is Your Life* – and he will remain the only lifeboatman to have appeared on Michael Parkinson's chat show.

Artist Tim Thompson depicts the moment before Coxswain Brian Bevan of the Humber makes a dash in alongside the *Revi* in the small hours of 14 February 1979. The coaster rolled over and sank five minutes after her four-man crew was rescued. (*Painting by Tim Thompson*)

12

HUMBER, 14 FEBRUARY 1979

Coxswain Brian Bevan of the Humber makes history as the only man to win Bronze, Silver and Gold Medals within the space of less than two months. His Gold Medal rescue is to a small Panamanian cargo ship, caught in a Force 9 and sinking 30 miles out in the North Sea. Bevan succeeds in taking off her four-man crew in massive seas with the skipper, the last to leave, hauled to safety minutes before the ship rolls over and sinks.

Those who volunteer to crew their local RNLI lifeboat soon discover the true meaning of the word commitment. It is not just that you are constantly at the beck and call of a pager, which will summon you to sea at the most inconvenient of moments and often for hours on end, you also need to be available for regular exercises, to attend training courses and to become involved in PR and fundraising exercises throughout the year.

If that is commitment to the cause of saving life at sea, how does one describe the sacrifices made by those who opt to become lifeboatmen at the Humber lifeboat station on the utterly remote tip of Spurn Point? With no other means of employment for miles around, every member of the crew there is full-time with the RNLI. Housing is provided for seven families and each crewmember works five days on, one day off, with every seventh weekend also off-duty. The nearest shop and village school is at Easington, 8 miles away, so wives and children of the crew are making almost as many sacrifices to the lifeboat service as their husbands.

There is no other RNLI station which is run in the same way, but its location, alongside the deep-water channel of the Humber Estuary with easy access to the North Sea, is so advantageous that it is well worth maintaining. The job of coxswain is even more important than usual here because it has to include the duties of the station manager, which means taking the decision to launch the lifeboat as well as commanding her at sea.

In 1975, Brian Bevan, who had begun his RNLI career as a volunteer crewmember at his home town of Bridlington, was made the new superintendent coxswain at Humber. His father had been in the Navy and then a Coastguard officer and Brian could not keep away from the sea as a boy. He used to skip school to go fishing and at the age of 12 spent six weeks away from home in Grimsby during the school holidays, working on a fishing boat. He, like so many fishermen, never learned to swim. He once recalled: 'I were chucked in at deep end when I were at school. I were frightened to death and I've never been in since. Besides, it's too bloody cold round here for swimming.'

He was only 28 when he was given the post, which made him the youngest coxswain in the country, but it would not be long before he was setting even more impressive records aboard the Humber lifeboat.

When he took up his post, the *City of Bradford III* was still on station. She was a conventional 9-knot wooden-hulled Watson class lifeboat, which had given sterling service at Spurn Point since 1954. But Humber was one of the busiest lifeboat stations and Brian Bevan knew that a lifeboat of the RNLI's most modern design, the Arun class, had already been allocated to the station. Although the introduction of the 14-knot Waveney class lifeboat in the late 1960s had moved the speed of all-weather lifesaving up a gear from the plodding 8 or 9 knots of conventional lifeboats, it was not until the Arun design was perfected and coming on stream in the mid-1970s that a true revolution occurred in the capabilities of the RNLI.

Here was a self-righting lifeboat, capable of 18 knots, with twin 460hp diesels offering lifeboat crews power and manoeuvrability unimaginable aboard a Watson class with her 50hp engines. The speed of the Arun meant that they were being called to emergencies much further out to sea than previously and with her watertight wheelhouse, they had the use of the most sophisticated electronic equipment to assist their search and rescue.

Thanks to an appeal by the Lord Mayor of Bradford, the Sheffield Lifeboat Fund and a donation from the International Transport Workers' Federation, *City of Bradford IV,* the seventh Arun class to be built, arrived at Spurn Point

Superintendent Coxswain Brian Bevan. (*RNLI*)

Brian Bevan and his crew aboard *City of Bradford IV*, soon after the *Revi* rescue in 1979.
(*Paul Berriff*)

in 1977. This was a time when the traditionalists were still somewhat sceptical
about the effectiveness of what appeared to them an awkward and oversized
vessel. One retired coxswain was heard to remark that it would be like taking
a block of flats to sea while others were uneasy about a lifeboat built out of
fibreglass – how could plastic be stronger than mahogany?

   If they wanted a young man capable of dispelling such doubts once and for all
by his actions aboard their new class of lifeboat, the RNLI made a sound choice
in placing an Arun under the command of Brian Bevan. Within the space of 46
days, in the winter of 1978/9, he carried out three outstanding services aboard
the *City of Bradford IV*, which earned him the unprecedented accolade of a
Bronze, Silver and Gold Medal, all presented at the same ceremony.

His Silver Medal came after he and his crew spent 12 hours at sea in an easterly gale and blizzards, going to the aid of a Dutch coaster, *Diana V*, whose cargo of maize had shifted and which was threatening to capsize in heavy seas, more than 70 miles out from Spurn Head. Bevan met the ship as she limped towards the Humber estuary and ran in alongside three times to take off all on board except the skipper. He stayed on board to save his ship, which he did, reaching the Humber with the lifeboat as escort in the early hours of 31 December 1978.

The Bronze Medal was won in equally hostile and wintry conditions only a day after Brian Bevan and his crew had survived the circumstances which had led to his Gold Medal award on St Valentine's Day 1979. Although utterly exhausted, the crew answered the distress call of a Romanian freighter, the *Savinesti*, with engine failure 37 miles to the south-east of Spurn Point. This time, although no one needed to come off the ship, the lifeboat spent 15 hours at sea standing by in mountainous conditions while the freighter limped back to the safety of the Humber.

What were the circumstances then, which barely 24 hours earlier persuaded the RNLI that Brian Bevan should receive the highest honour for saving life at sea, and his crew, Second Coxswain Dennis Bailey, Mechanic Bill Sayers, Assistant Mechanic Ronald Sayers, Dennis Bailey, Jnr, Peter Jordan, Sydney Rollinson and Michael Storey, the Bronze Medal?

A north-easterly gale, gusting to Force 9 and increasing; snowstorms; midnight. It is difficult for any east-coast lifeboatman to sleep easily when the weather is in that sort of mood at that time of year. The last thing they want is a call, but somehow they expect one.

It is 13 February 1979. Brian Bevan is woken three minutes before midnight by the Coastguard. A small cargo vessel, the *Revi*, Panamanian registered and carrying silver sand from France to Newcastle, is in distress 30 miles north-east of Spurn lightvessel. Heavy seas have ripped away her hatch covers and water is entering her hold.

Brian's wife Ann is drowsily aware of the sound of the gale outside, but she has never seen her husband frightened and has great confidence in his skill at the helm. 'Tara love,' she murmurs and goes back to sleep.

Eighteen minutes later and eight men, including the coxswain, are aboard the 54ft Arun class lifeboat *City of Bradford IV*. She has slipped her mooring in the shelter of Spurn Point and is heading into the darkness out to sea at full speed.

Now she is clear of the Humber, climbing and falling from mountainous head seas. The impact into the trough of one 20ft sea opens every electric circuit breaker and plunges the wheelhouse into darkness. The coxswain reduces speed to 14 knots, the waves increase, some as high as 35ft.

The lifeboat has made 50 minutes of gruelling progress when the *Revi* puts out an even more urgent distress: she is now slowly sinking. She requests the British ship *Deepstone*, which is already standing by, to stand in close to. The lifeboat still has 8 miles to run.

The lifeboat crew get their first glimpse of the casualty at 1.36am. Her master, in a desperate attempt to reach the River Humber, is steaming at full speed towards them, his ship continually buried by the huge seas. The wind is at Storm Force 10 as the lifeboat takes up station close astern of the *Revi*.

A few minutes pass, then the master radios that he is slowing down. He wants two crewmembers taken off. Coxswain Bevan asks him to stop so that he can work out the best way to carry out this apparently impossible request. He tells the master to steer south at slow speed and to have the two men on the boat deck on her starboard bow, ready to jump.

Then the lifeboat moves in, fendered on the port bow and her crew ready on the foredeck, their lifelines secure. The lifeboat edges in under the casualty's starboard quarter and a huge wave crashes over the port quarter, completely engulfing the coaster's stern.

Brian Bevan throttles full astern just in time to see *Revi*'s bulk fall back down, missing his foredeck by inches. Again and again the lifeboat makes an approach and is forced back, the casualty often towering 20ft above the heads of the men on the foredeck. At last the right moment comes and the two crewmen are able to throw themselves into the waiting arms of the lifeboatmen.

With two of his men safe, *Revi*'s skipper hopes to continue the desperate run for the River Humber, but it is not to be. Only five minutes pass before the accommodation begins to flood, the cargo of sand shifts and the ship is listing 45 degrees to port. Now he and the mate must abandon ship.

His final act at the helm is to turn the bow to the west to give a lee on the low port side. Meanwhile the lifeboat crew are fighting to secure fenders, this time to the starboard side, and to make themselves fast to the pulpit rails. The *Revi* is clearly sinking, she is down by the bow and seas are sweeping clear across her full length.

A massive sea breaking clean over both vessels forces the lifeboat contemptuously away from the side of the casualty as Brian Bevan makes his first approach. Unflinching, he comes in again; another thunderous wave hurls the lifeboat aside. Only on the 12th attempt does the sea provide sufficient respite for the mate to jump 6ft from his position on the ship's port quarter. He lands in the arms of the lifeboat crew who break his fall and hurry him below.

The *Revi* is now at a crazy angle; her bow is below the waves and her stern juts clear of the water, menacing the lifeboat with a lethal blow if she dares to come close. Her master is praying that the coxswain will dare. He is hanging on for dear life to the outside of the stern rails, ready to jump. Nine times the lifeboat gets close but not close enough. On the tenth attempt the stricken vessel's stern suddenly soars 20ft clear of the water and then plummets towards the lifeboat's foredeck and the crew immediately beneath.

Only the coxswain's lightning reaction – ramming the throttles full astern – and the power of the Arun class engines avert a total tragedy, literally by inches. Then three successive seas cover the *Revi* completely. The lifeboat crew cannot believe their eyes as the water clears and the captain is still seen hanging on to the stern rails. But the ship is about to roll over. Bevan decides on a dash in to the casualty in a trough between two waves. The lifeboat drives in under the port quarter, strikes the stern, but the captain is able to jump. He lands on the very edge of the lifeboat's deck and is only prevented from being lost overboard by the strong arms of the crew. Five minutes later the *Revi* rolls over and sinks.

The four dazed and bedraggled survivors are eventually put ashore at Grimsby and taken into the care of the Mission for Seamen. Bevan and his crew snatch a bite to eat and then sail back across the estuary to their mooring where they arrive at 7.20am.

Reflecting on the incident several days later, Bevan recalled the moment the *Revi's* captain made it safely into the wheelhouse of the lifeboat:

He'd taken his boots off and he'd cut his foot but he was calm as anything. He said, 'Are you the skipper?' I said, 'Yes.' He said, 'Thank you very much,' and we shook hands.

Brian Bevan and his wife Ann, in London for the presentation of his Bronze, Silver and Gold Medals, all earned within 46 days in the winter of 1978/9. (*Maggie Murray*)

You feel very chuffed when you do a job like that, when normal ships are running for shelter and you're going out.

He then went on to consider what had led to their success, much of which was patience, waiting for the right moment to make a run in and the right moment to grab a man to safety. He also praised his crew:

Nights like that sort out the men from the boys. The success of that rescue was the crew. There were eight of us because the second cox's son had come with us and if there'd been a weak link in the chain we couldn't have done it. Nights like that you have a full-time job watching yourself without having to keep an eye on a weak link. We're all pretty close.

Penlee's lifeboat *Solomon Browne* at sea in Mount's Bay, off the Cornish coast, not long before her launch to the *Union Star* in December 1981. (*Norman Fitkin*)

13

PENLEE, 19 DECEMBER 1981

Trevelyan Richards, coxswain of Penlee lifeboat, is lost with his seven-man crew and all those he was attempting to save from the coaster *Union Star*, driven by a hurricane up against a jagged Cornish cliff-face. This heroic rescue attempt, which earned Richards the RNLI Gold Medal posthumously, had a profound effect, not just on the village of Mousehole where the coxswain and crew had lived, but across the RNLI and the nation.

For most people in their 20s and even early 30s, the name Penlee will mean nothing. It sounds Cornish, but search for it on a map and you will find neither village nor settlement with that name. For those who can remember some of the events of the 1980s, the name may well conjure up a tragic story about eight brave men, which dominated the news in the days before and after the Christmas of 1981.

Penlee lifeboat takes its name from an insignificant rocky point under the toe of the Cornish peninsula, just beneath the coastal road that runs between Newlyn and the huddled village of Mousehole. Nowadays, Penlee lifeboat is a powerful Severn class which lies afloat in the busy fishing harbour of Newlyn, but in 1981 a 47ft Watson class lifeboat, *Solomon Browne*, sat in her boathouse at the top of a slipway, built into the rocks of Penlee Point. The boathouse still stands today, just as it was at the moment the lifeboat launched for the last time into the howling night of Saturday 19 December 1981, deliberately and hauntingly preserved as a memorial to her crew.

Earlier on that Saturday, Mick Moreton, the 32-year-old master of the newly built 1400-ton coaster *Union Star*, had seen his well-laid plans begin to come unstuck. Maybe he had not gained permission from his owners, but no one would be any the worse off if he made a slight detour in his ship's maiden passage from Holland to the Irish Republic to pick up his wife and two teenage step-daughters from Brightlingsea in Essex. It would add about seven hours to the Ijmuiden to Arklow run, but he would still be able to deliver his cargo of fertiliser in time and it would give his family a pre-Christmas treat.

The treat began to turn sour when the weather took a rapid turn for the worse as the coaster progressed down the English Channel. Soon she was corkscrewing through heavy seas, whipped up by gale-force winds from the south. The forecast was that these would increase to Force 10 before any improvement could be expected. They were conditions that would certainly put the new ship through her paces, even if they would also convince the skipper's wife and family never to hitch a ride with him again.

Then, as darkness fell, the persistent throb of the *Union Star*'s engine suddenly ceased. Lights flickered and went out, to be replaced by the dim glow of emergency lighting. The only sound was of the wind howling through the rigging and the hull churning in the swell as the vessel rolled drunkenly, broadside on to the weather.

Ships sometimes break down at sea and can drift for long periods while their engineers make the necessary repairs without it causing undue anxiety aboard. Here, though, however much Mick Moreton was determined to show calm and confidence, he realised that some extra precautions were necessary. At 6.04pm he called up the Coastguard at Falmouth to tell them of his engine problem and his position, which by his reckoning – his radar had gone out of action with the power failure – was 8 miles east of Wolf Rock. This would put him about 6 miles off the south coast of the western tip of Cornwall and he would be drifting towards it at a rate which was very difficult to gauge. He asked that a helicopter could be made ready, in case the situation deteriorated.

In the course of the next 15 minutes Falmouth Coastguard contacted HMS Culdrose where a Sea King helicopter was prepared for action; they also asked the Penlee lifeboat coxswain to anticipate a launch and they radioed the salvage tug *Noord Holland*, which was moored in Mount's Bay, off Penzance. It was not an easy situation for the Coastguard officers that night. The message from

Solomon Browne at the top of her slipway at Penlee Point. (*Andrew Besley*)

the *Union Star* had only been precautionary and they had been given no idea how quickly her engine could be fixed. They had to rely on the skipper to judge the urgency of his predicament and he was still sounding calm and in control and seemed far from issuing a Pan broadcast (i.e. 'Urgent help required but not in imminent danger'), let alone a full Mayday. He had also refused a tow from the *Noord Holland*, fearing a large bill for his owners which still might not be necessary.

But with the ferocious weather that night, the Coastguards were extremely concerned. A sector officer was sent to the Land's End station to try to locate the *Union Star* on the radar. It soon became clear that she was drifting towards the shore at a far greater rate than her skipper had supposed. By 7pm she was barely 3½ miles from the coast. At roughly the same time, her engineer had diagnosed the engine problem: seawater had somehow got into the fuel supply. Even in calm conditions it would have taken a good two hours to remedy. Now, hurled about the engine-room and in semi-darkness, a miracle was needed for the task to be accomplished in twice that time.

Finally, at 7.20pm, Moreton agreed to a Pan broadcast and the tug, *Noord Holland*, announced she was on her way to the *Union Star*, ETA two hours,

The old Penlee boathouse, which has stood empty since the night of 19 December 1981. (*Nicholas Leach*)

a suitable salvage agreement having been reached between the two vessels' owners. Quicker on the scene would be the Royal Navy helicopter R80, which had been asked to scramble with the primary task of taking off the skipper's wife and her children. The Coastguard district controller, Robbie Roberts, still delayed calling out the lifeboat; a Mayday had not been issued and no one wanted to send the crew out on a night like that unless it was absolutely necessary.

And it was a terrible night. Guy Buurman, captain of the *Noord Holland*, could make no more than 4 knots into 15–20ft seas, which he later said, 'for the combination of size and steepness were the worst I have ever seen'. At the controls of the rescue helicopter, Lt Cdr Russell Smith, an American on exchange from the US Navy, encountered 80-knot winds as he headed across Mount's Bay. Below him he could see waves up to 40ft and his aircraft was being hit by sea-spray despite its altitude of 400ft.

It was only when the helicopter located the *Union Star* that everyone realised how little time there was left to save the ship, or, if not the ship, at least her five-man crew and three passengers. Learning her true position – just 2 miles off the shore opposite Tater Dhu light – District Controller Roberts quickly calculated that only about an hour and a quarter remained before she would be on the rocks. The tug was still a good hour away and, by the time she arrived, the chances of her passing a tow so close in to the shore were negligible. The people on board the *Union Star* probably only had two chances now: the helicopter or the lifeboat. But the lifeboat was still in her shed and her crew only aware that they might be needed and therefore still enjoying their Saturday night a mile away in Mousehole. It was 7.50pm when Coxswain Trevelyan Richards received the call from the Coastguard, asking him to launch without delay.

The boathouse was alive with activity in minutes. Trevelyan Richards, known to everyone locally as Charlie, was among the first to arrive. He was 56 and a bachelor, living with his elderly mother. He commanded a great deal of respect from his crew for his unrivalled experience and skill, both as a trawler skipper and as lifeboatman for more than 30 years. Six years earlier he had been awarded the RNLI Bronze Medal after a horrendous day at sea in a hurricane hoping to rescue a crew who had abandoned ship 24 miles out to sea. Sadly there had been only bodies to recover.

Coxswain Trevelyan Richards.

Mechanic/Second Coxswain Stephen Madron.

Mechanic/Second Coxswain Stephen Madron had, like his coxswain, driven straight from home. He was 33, married, with a young son and daughter, and had family connections with the lifeboat going back generations. His grandfather Edwin had been coxswain at Penlee and won the Silver Medal for bravery in 1947. As well as his mechanic's job, Stephen worked as boatman to the local pilots.

Mechanic Nigel Brockman. (*RNLI*)

His assistant mechanic, Nigel Brockman, 43, had been at the Mousehole British Legion, watching the ladies' darts match as he waited to hear whether the lifeboat would be needed. He was a fish salesman in Newlyn, having also been a fisherman, and had served on the lifeboat for 16 years. He was married and one of his three sons, Neil, 17, only recently enlisted as a crewmember, was also at the boathouse to see if he was needed.

John Blewett.

John Blewett, 43, had not allowed a heavy cold or his daughter's 15th birthday party, which was in full swing, to stop him from leaving the house and family and hurrying to the lifeboat station. Another Mousehole man, born and bred, he worked as an engineer for British Telecom and volunteered his expertise annually to wire up the famous Christmas lights in the harbour.

Another true local and long-standing crewmember was Barrie Torrie, 33, a fisherman, married, with two young sons. He had just settled down to watch a film, *The Lost City of Atlantis*, on television when the call came.

Kevin Smith, 23, had moved to Mousehole with his family from Yorkshire when he was young. Now a seaman aboard the Cunard vessel *Samaria*, he had been at home on leave, watching the same film as his colleague and nursing the after-effects of viral pneumonia until he, too, received the telephone summons.

Barrie Torrie.

Kevin Smith.

Gary Wallis, 22, had been at the Legion with Nigel Brockman and had been given a lift by him to the boathouse. He, too, had moved to Cornwall with his family who were from London originally, and he still lived with them in the village. He worked as a fisherman out of Newlyn.

All these men, apart from the young Neil Brockman, were regular members of the crew and were soon togged up in foul-weather gear and lifejackets and making ready aboard the lifeboat. Coxswain Richards now surveyed the less regular members of the crew who had come at the sound of the maroons, weighing up whom to take as his eighth crewman. Some he thought too old for what he suspected lay ahead and one keen volunteer, Neil Brockman, he turned down saying, 'No more than one from a family on a night like this.'

As those around absorbed this sobering pronouncement, another man burst into the boathouse. It was the landlord of The Ship Inn in Mousehole, Charlie Greenhaugh, who had managed to extricate himself from one of the busiest nights of the year behind the bar. Richards was very happy to take him as his eighth man. He was 46 and had served in the Royal Navy and the Merchant Navy and was a popular figure in Mousehole, having been given the honour of switching on the Christmas lights the day before.

Even to launch the lifeboat took tremendous courage that night. Great waves were sweeping the slipway as the *Solomon Browne* was inched forward out of the boathouse under her own

Gary Wallis.

Charles Greenhaugh.

weight on the winch wire. There was a great risk she would be swept away while still on the slipway but, timing her release with utter precision, the coxswain allowed the lifeboat a full run down into a deep trough, giving her maximum momentum to power away from the rocky shore.

Meanwhile, out at sea, an extremely dangerous operation was under way. The Royal Navy helicopter R80 was hovering as low as the pilot dared over the *Union Star* in an attempt to lower the winchman to the cramped after deck so as to take off the three female passengers. It was impossible to maintain a steady hover in the violent gusts of wind and the winchman was swinging alarmingly above the coaster which rose and fell beneath him and whose mast whipped menacingly back and forth through nearly 180 degrees.

Lower and lower the helicopter came, trying to reduce the swing on the winch wire, but after the third attempt, when the mast had missed striking the rotor blades by only 10ft, the pilot knew he would have to try another method. This involved dangling a 'high line' from the helicopter which could be grabbed by someone on deck and which the winchman could then use to guide himself down to the ship. But the wind and violent motion of the ship thwarted these efforts, too. Once a crewmember did succeed in grabbing the line, but then the coaster plunged into a trough and he lost his hold on it. Then, for a heart-stopping moment, the line became entangled in the mast; fortunately, the winchman was able to free it from his mid-air position but, for the second time, the helicopter had come within an ace of crashing on to the deck of the *Union Star*. A brief radio exchange between Lt Cdr Russell Smith and Mick Moreton followed:

R80 to *Union Star*: 'Too difficult for us as far as safety is concerned. We're getting very close to your mast and we don't have a long enough line.'

Union Star to R80: 'OK. Very much obliged for your assistance. Going to put an anchor down.'

This was more easily said than done, but one of the crewmembers did manage to get forward to release the anchors, risking, as he did so, being swept over the side by massive waves that broke clean over the hatch covers. The starboard cable snapped almost immediately the anchor reached the seabed. The port anchor was then let go and although it did not hold firm, its drag slowed the coaster's drift towards the rocks. It also had the effect of turning the bow towards the oncoming seas, which would often immerse the forward part of the ship and then crash against the bridge with terrifying force.

By now, at 8.43pm, both tug and lifeboat were arriving on the scene. The *Union Star*'s engineer had just managed to restart one of the auxiliary generators which gave the ship electric power again and she was well lit. It did not take Captain Buurman, the tug skipper, long to realise that the *Union Star* was far too close to the shore for him to manoeuvre safely to connect a towline. In any case, it was now out of the question to send one of the coaster's crew forward to attach the line; he would have been swept away instantly. The tug could only stand off and watch the lights of the helicopter and the lifeboat as they moved in closer.

The lifeboat had taken half an hour to travel the 2½ miles from her station. It would have been one of the most alarming passages any man aboard her had ever made. Waves were 35 or even 40ft high, the wind was gusting to hurricane force. Only the ablest, strongest helmsman could have kept the lifeboat from capsizing in those seas. For the first time, Trevelyan Richards could be heard speaking to the *Union Star* over the radio:

Penlee lifeboat to *Union Star*: 'Understand you had trouble with the chopper. Do you want for us to come alongside and take the woman and children, over?'

Union Star to Penlee lifeboat: 'Yes please. The helicopter is having a bit of difficulty getting to us, so if you could pop out, I'll be very much obliged, over.'

The helicopter crew did, in fact, make another attempt to trail a high line to the ship after this brief exchange. They had attached a weighted bag at the end of the line, but when, at one point, it dipped in the water, a breaking wave caught it and tore the bag from the line which then streamed away uselessly in the wind.

The lifeboat crew would have realised now that they were the only people who had a chance of saving any lives. Advising everybody to get ready to come off, the coxswain made an approach along the coaster's port side, his bow to seaward. The view from the helicopter, lighting the scene with spotlights from above, was of the lifeboat colliding with some force against the ship's side and of her crewmembers throwing ropes over the rail to keep themselves alongside for as long as possible.

Others waved furiously to the people on the bridge to make a run for it, but then a huge wave reared up ahead and the lifeboat was forced to pull away to save herself.

Coxswain Richards' next approach was from the seaward side, drifting down towards the after end of the *Union Star* and holding on with ropes as long as he dared. Still no one emerged from the bridge. Again and again the same approach was tried; any weaker vessel would have long since cracked open with the force with which the lifeboat continually struck the metal plates of the heaving coaster. On one approach one of the lifeboat crew was seen attempting to board the *Union Star*, presumably to try to drag people physically to safety. He was pulled back, however, either by the violent motion or by his fellow crewmembers. On another occasion the lifeboat's bow was lifted over the ship's rail and landed with such impact on the side decking that the masthead light went out. In moments she was back in the water, her coxswain pulling away to make yet another attempt.

By 9.20pm, the danger to the coaster, lifeboat and helicopter had become extreme. They were no more than 200 yards from the cliff. Rocks and ledges lay just beneath the surface of the water, exposing themselves momentarily in the wake of the gigantic seas that passed over them. The 300ft cliffs loomed up unseen behind the helicopter, hovering at only 100ft; the slightest contact with them would have been fatal for her crew who could feel the 100mph gusts blowing them backwards involuntarily. There were barely ten minutes left before the *Union Star* would be on the rocks.

Russell Smith, at the controls of the helicopter, decided he could not stand by watching the lifeboat risking everything on each approach without having one last attempt himself at putting his winchman down to the ship. The first time he went in the flailing mast missed his rotors by no more than 5ft. On the second try it came closer still and in the desperate avoiding action, the aircraft was flung back 30 or 40 yards by the wind and the winchman, Steven Marlow, swung upwards on his line and struck the underside of the helicopter. The pilot regained control and the winchman was hauled back on board, bruised but still alive. There was nothing more they could do but observe.

To their considerable disbelief, the lifeboat below them continued with the same persistent pattern of approach, even though no one stirred from the bridge. Then the anchor chain parted. Immediately the ship slewed round, broadside to the waves and she was swept, threatening to capsize at any moment, towards the breakers on the shore. The lifeboat went after her. The first attempt to get close came to nothing; on the second approach a wave caught her and carried her right up on to the hatches of the *Union Star*. For a moment the lifeboat stayed there like some incongruous deck cargo, then the ship rolled and the *Solomon Browne* slid back into the sea stern-first.

Even that near disaster was not enough to deter the Penlee crew from trying again. This time, at long last, the eight people aboard the *Union Star* realised that their only hope was to get to the rails on the lower side deck. Eight figures, clad in orange lifejackets, tumbled down the steps from the bridge and clung on to the railings, fighting to stay on board as seas washed over them and the ship played havoc with their balance. Suddenly, the *Solomon Browne* loomed up out of the spray and darkness and thumped against the side. Arms reached out to grab anyone within reach. When at last the lifeboat had to pull away, the helicopter crew could see that they had some survivors aboard but that there were two people still on the coaster and one, if not two, in the water.

Then, moments later, at 9.21pm, they heard over the radio:

Penlee lifeboat to Falmouth Coastguard: 'Falmouth Coastguard, this is Penlee lifeboat, Penlee lifeboat calling Falmouth Coastguard.'

Falmouth Coastguard to Penlee lifeboat: 'Falmouth Coastguard, Penlee lifeboat, go.'

Penlee lifeboat to Falmouth Coastguard: 'We got four men off – look, er hang on – we got four off at the moment, er . . . male and female. There's two left on board . . .'

The message ended abruptly with a crashing noise. But Russell Smith could see the lifeboat still, apparently under control and heading out to sea. He took

One of the fragments of the *Solomon Browne* washed up at Lamorna Cove, close to the site of the disaster.

The *Union Star* photographed at first light on 20 December 1981.

this as his cue finally to lift his aircraft out of the death trap where she had been hovering for so long and head back to Culdrose. He and his crew had been operating under extreme duress for two-and-a-half hours and any more time airborne could have proved fatal. He had assumed the lifeboat had made the same decision to turn for home.

At this point there was only one other witness left, the tug *Noord Holland* standing off about a mile out to sea. Her skipper, Guy Buurman, listening to the vain attempts by the Coastguard to regain radio contact with the lifeboat and watching intently from his wheelhouse, could see the *Union Star* right up close to the cliff and, intermittently, the lights of the lifeboat. His last view of the lifeboat was when she appeared high on the crest of a wave, silhouetted against the coaster's lights. Minutes later, the ship suddenly went dark, the moment she was at last tumbled over at the foot of the cliffs.

No one will ever know what calamity overcame the *Solomon Browne* and her crew. By the time cliff rescue teams arrived at the scene, the *Union Star* was already wrecked at the foot of the cliffs and there was no sign of the lifeboat. Did she turn back again to try to rescue the others? Did she lose power or control at a vital moment close to the shore? Had she been badly damaged on one of her earlier approaches to the *Union Star*? The wreckage gave no real clues other than that she was ultimately subjected to the most shattering and violent force imaginable. Never before has a lifeboat been smashed to smithereens in the way that had happened here. The largest portion of the boat, including the heavy engine compartment, was found 300 yards to the east of the *Union Star* which suggests she met her fate here.

Some of the victims' bodies were never found. Sympathy with the grief felt by all their families, particularly the lifeboat crew's, was universal across the UK and Ireland. The story of the disaster and its aftermath amply sustained the news media throughout the slack Christmas and New Year period. Spontaneous donations poured in for the Mousehole families who became unwilling objects of fascination with the press, not least because of the eventual size of the fund in their favour.

Everyone recognised that Trevelyan Richards and his crew (each of whom received a posthumous Bronze Medal, to accompany their coxswain's Gold) had paid the ultimate price of volunteering for the RNLI. The public inquiry into the disaster found that nobody could or should be blamed for their death, but supporters became all the more determined to ensure that crews were given the best possible tools for their selfless work. This positive public attitude, reawakened by the Penlee disaster, probably helped the RNLI more than anything else in the 1980s to raise the money needed to accelerate the modernisation of its fleet. The new Penlee crew, which had re-formed, manning a relief lifeboat only days after the disaster, received a new Arun class, the *Mabel Alice*, in 1983. They and many other lifeboat crews around the UK and Ireland would have had considerably longer to wait for a more capable lifeboat without the tragedy which overtook the eight men aboard the *Solomon Browne*.

Trevelyan Richards's mother Mary, with her son's posthumous Gold Medal, alongside Coxswain Mike Scales of Guernsey, who also received a Gold Medal at the same ceremony for a rescue a week before the Penlee disaster. (*Guernsey Press Co. Ltd*)

Neil Brockman, son of
the mechanic, Nigel, who
volunteered, but was not
taken on the night of the
Union Star rescue. He
later became coxswain of
Penlee lifeboat. (*Maggie
Murray*)

One of the first people to step forward to be part of the new Penlee crew
following the disaster was the 17-year-old Neil Brockman, son of the late
assistant mechanic. He went on to become coxswain, a position he held until
2008. In 1995 he was awarded the RNLI Bronze Medal for his part in the
rescue of five men aboard a trawler off Land's End. It was a proud day for him,
his family and the people of Mousehole, pointing up the indomitable spirit of
Cornish seafaring blood, and reminding the world that triumph was as much a
part of Penlee lifesaving as tragedy.

The sail-training ketch *Donald Searle* among 20ft breaking seas off Hayling Island and dragging her anchors ever closer to the shallows, October 1992. (*RNLI*)

14

HAYLING ISLAND, 25 OCTOBER 1992

Rod James and Frank Dunster earn the RNLI Silver Medal taking crewmembers off the 75ft ketch *Donald Searle*, aground in heavy seas at the entrance to Chichester Harbour. James becomes the only man to date to win a second Silver Medal aboard an inshore lifeboat.

It is a wild October Sunday morning and it reminds Frank Dunster of the weekend on Hayling Island more than a decade earlier when he had spent a frantic three-and-a-half hours at the helm of the Atlantic 21, attending to eight separate incidents inside and outside Chichester Harbour where pleasure-seekers had been caught out in a Force 9 south-easterly gale.

Waves had been up to 8ft high even inside the harbour and, beyond its protection, the 21ft rigid inflatable lifeboat had come close to capsize on a number of occasions, once being thrown almost beyond the vertical by a heavy sea. One of his crewmembers, Rod James, had twice gone overboard that day, once by accident when helping to get a grounded yacht off the sands and once on purpose when he braved huge breakers to grab a 17-year-old who had been clinging to a groyne for his life and who had just let go.

For saving the boy's life (a pupil at the school where he taught), James became only the third man from an Atlantic 21 to win the RNLI Silver Medal; for his helmsmanship, Frank Dunster won a bar to an earlier Bronze Medal.

All that happened on a busy Saturday in mid-September 1981 and at least today, being later in the season, there is a bit less activity on the water. The 50-knot westerly gale has not stopped the windsurfers in the harbour, though; rather it has encouraged many of the hardier ones and Frank Dunster is not particularly surprised to hear, on arriving at the lifeboathouse, that the lifeboat is already out, bringing one of them in difficulties to safety.

Just before midday one of the small group of people gathered in the empty lifeboathouse holds up a hand to silence the others. He is closest to the radio receiver and is listening to an exchange between the Coastguard and a yacht skipper. It is what he thought he heard – '*Mayday, Mayday, Mayday*' – the *Donald Searle*, a 75ft sail-training ketch, has got her anchors down at the eastern end of Chichester Bar, her sails have blown out and her engines have overheated and failed. Twenty-feet breaking seas are hurling her about and she is dragging slowly eastward towards the sunken Target Wreck. There are 17 crew and trainees on board.

The Coastguard is quick to react. The 47ft Tyne class lifeboat from Bembridge is called and they also notify Rod James, today's helmsman aboard Hayling Island's Atlantic 21, *Aldershot*. They realise, though, that he is not instantly available as he is still dealing with the windsurfer.

Frank Dunster cannot idly stand by, knowing how dangerous a position the yacht is in. He has a 28ft, single-engine, rigid inflatable boat (RIB) of his own, *Hayling Rescue*, kept at the nearby marina and used often as a rescue boat for the local sailing club. With two volunteers from the lifeboat station, Damien Taylor and Evan Lamperd, he is soon afloat and powering south towards the harbour entrance with the jagged outline of 20ft breakers on the bar ahead of him.

He reaches the bar and realises immediately that there is no hope of the Bembridge lifeboat getting near enough to the ketch to tow her clear of danger. She is in far too shallow water and dragging her anchor perilously close to the wreck. They must get the crew off. He radios urgently for helicopter assistance and steels himself and his crew for an attempt to get alongside. Whether he is in the lifeboat or his own rescue boat, there is one type of sea that he always fears. It is the 16ft solid wall of water which becomes unstable in the shallows, with the top 4ft curling, then breaking. You can't get clear of it so you have to face it head-on. There is no knowing whether the boat will go

Frank Dunster, holder of the Silver and two Bronze Medals for bravery. (*RNLI*)

Frank Dunster at the helm of his RIB, *Hayling Rescue*.

through it or ride up it and possibly fall over backwards with the force of the breaking crest.

Using all his experience and skill, accelerating, then slowing as he reads the mood of each wave, Dunster draws near to the yacht. Just as he feared, one wave throws his bow high into the air, but it slams down again, much to the relief of the crew. It takes two attempts to get alongside the casualty's starboard quarter, but when they do, one of the yacht's crew is ready and is pulled aboard *Hayling Rescue*. The second approach nearly ends in disaster. First, the ketch surges high above her on a wave, then crashes down, inches from her bow. Then a female crewmember leaps from the *Donald Searle*, misses the RIB and lands in the water. Taylor and Lamperd are quick to reach over the rubber sponson and haul her, spluttering, into the boat.

Dunster knows he must not chance his luck any further. One of these waves would surely capsize his boat and with two survivors now on board, his job is to get them to safety before that happens. Bembridge lifeboat and the helicopter could not be far away now so, with that thought, he turns north towards the shore, this time fighting to keep the boat straight as she rides the heavy surf charging in haphazardly through the harbour entrance. He reaches the lifeboat station at 12.35pm, only 23 minutes since he set out, although it feels more like a lifetime.

By now, Rod James aboard the Atlantic 21 lifeboat has landed the shivering windsurfer and, with his crew, Christopher Reed and Warren Hayles, is taking his turn at negotiating the brutish walls of water on the bar. One is so steep that it stands the lifeboat up on its end. Warren Hayles is thrown backwards out of his seat, but manages to stay on board. James's grip on the steering wheel keeps him on board, although he is convinced the boat is going over on top of him and he takes a deep breath in anticipation. The bow teeters at the vertical and both engines cut out.

People on the shore gasp as they see the bow reaching for the sky and will it to return the way it went up. Their prayers are answered and James and Reed, still in their seats, stretch forward to press the port and starboard starter buttons. The engines fire up just before the next mammoth wave arrives and the helmsman has control once more. Elated to have survived such a near disaster, James is now all the more determined to get to the stricken ketch.

Double Silver medallist Rod James, at work as a schoolteacher.

As he manoeuvres close to her, he has to decide: are the people aboard better off staying put or should he try to get them ashore? They will not be particularly safe aboard the lifeboat, but if the ketch founders, he will have 15 people in the water and that would be much worse. The pilot of the Sea King search and rescue helicopter from Lee-on-Solent, which has just arrived, clearly believes that evacuation is a must, as he is attempting to pass a line to the vessel's crew in preparation for winching.

James drives the lifeboat's inflated sponson up against the ketch's side five times, staying there just long enough each time for Christopher Reed, in the bow,

to grab a crewmember before the larger vessel rises up on a swell and falls back on top of where the lifeboat has just been. The Atlantic 21 helmsman then sees that the helicopter has still not got its line to the remaining yacht crewmembers – it is being hampered by the violent motion and the mizzen mast. On his next

The view from the Hayling Island Atlantic 21 inshore lifeboat of the stricken ketch, *Donald Searle*, the Coastguard helicopter and Frank Dunster's RIB, *Hayling Rescue*.

run in he asks Reed to leap aboard to help, but not before two more yachtsmen are dragged into the lifeboat.

Reed wedges himself between the aft cabin and the guard rail and grabs the line from the helicopter to allow the winchman to make his descent. Leaving

Safe in the boathouse, two young survivors from the sail-training ketch *Donald Searle*. (*Zac Austin*)

Reed aboard the ketch, Rod James now heads for Hayling lifeboat station to land his seven survivors knowing that Bembridge lifeboat, now on scene, and the helicopter can complete the evacuation. On the 15-minute journey back to the shore, the Atlantic 21 passes Frank Dunster aboard *Hayling Rescue* on his way back out to the rescue scene.

Of the remaining survivors, Bembridge lifeboat takes one aboard, but both vessels are damaged in the process when the yacht is thrown 20ft to leeward by a sea. The remaining seven, together with Christopher Reed, are hoisted into the helicopter and landed near the lifeboat station on Hayling Island. Both *Hayling*

Rescue and the Atlantic 21, which is also back at the casualty after landing survivors, can now, once more, return to station and safety.

The empty *Donald Searle* is recovered the next day, a few hundred yards west of the Target Wreck and severely damaged by the pounding she has received on the sands.

Both Rod James and Frank Dunster received the RNLI Silver Medal for their achievements in conditions many would have judged too extreme for the boats at their disposal. Rod James is still, to this day, the only inshore lifeboatman to win two Silver Medals and Frank Dunster joined a very rare breed to receive three bravery medals in his career as an inshore helmsman.

The freighter *Green Lily* aground and at the mercy of the sea, off the Shetland Island of Bressay, November 1997. (*RNLI*)

15

LERWICK, 19 NOVEMBER 1997

In a rescue in which helicopter winchman Billy Deacon loses his life after saving the last ten men from the freighter *Green Lily* in a near hurricane, Coxswain Hewitt Clark earns the RNLI Gold Medal for bringing his lifeboat alongside the ship within yards of the shore and saving five of her crew.

What circumstances could lead to a soft-spoken, modest, Shetland harbour pilot boat coxwain finding himself as a travelling guest of honour aboard the liner *Queen Elizabeth 2*, sharing the speaker's platform with the likes of Terry Waite and the author Bernard Cornwell? Whatever they were, they also caused Hewitt Clark to be the subject of an ambush by a man whose face was familiar and who was clutching a red book with gold lettering on its cover and who would later recount his life's exploits to a BBC audience of millions.

There was quite some story to tell; for his services as coxswain of Lerwick lifeboat, Hewitt Clark, MBE, had four times received the RNLI's official Thanks of the Institution on Vellum, the Bronze Medal for bravery three times, the Silver once and, in November 1997, the ultimate recognition of a Gold Medal when five petrified Croatian seamen were dragged aboard his lifeboat while 50ft seas broke over them. Hewitt Clark was the most decorated lifeboatman of his time and whether he was comfortable about it or not, people wanted to listen to him.

The Shetland Islands may be remote, but during Hewitt Clark's tenure as coxswain they would have seen a greater weight and variety of seagoing commercial traffic than most other parts of the British Isles. The North Sea

Coxswain Hewitt Clark, who earned the RNLI Gold Medal for his participation in the rescue of the crew of the *Green Lily*. (*Maggie Murray*)

oil and gas industry with its base at Sullom Voe reached its peak of activity between the 1970s and 1990s and the decrepit but abundant Russian fish factory ships of that era added to the regular island traffic of ferries, fishing boats and summertime pleasure craft.

The factory ships, or 'klondikers', were a particular source of concern for the rescue services. Poorly maintained and crowded with workers employed in the process of canning mackerel caught by local fishermen, they were especially vulnerable when the weather blew up in the area. The service records of Lerwick lifeboat are littered with incidents where these ships, with names such as *Azu*, *Lunokhods 1*, *Borodinskoye Polye* and *Pionersk*, had been driven ashore by a gale and the lifeboat had had to take considerable risks close to the rocks to rescue the people on board.

It was to the *Pionersk* in October 1994 that Hewitt Clark earned his Silver Medal. The factory ship had run aground 3 miles south of Lerwick in storm-force winds. There was less than 100ft between the wrecked ship and the cliff, but this did not deter Clark from taking the Arun class lifeboat *Soldian* into that gap of shallow water, in darkness and confused breaking seas, to allow 64 people to crowd aboard the lifeboat and be brought safely ashore.

Three years after the *Pionersk* service, Lerwick was to say goodbye to its Arun class lifeboat *Soldian*, which had served her crew so reliably for 19 years. In her place came a lifeboat which, to the untrained eye, looked similar in many respects, although she was clearly bulkier than her predecessor. She was a Severn class lifeboat, one of a new generation of all-weather lifeboat design and Lerwick was one of the earliest stations to be allocated such a boat. At a length of 17m, the class was the largest in the fleet and had been designed to achieve a maximum speed of 25 knots, making her the most powerful RNLI lifeboat ever built.

Getting a lifeboat to be strong enough for the task, yet light and powerful enough for the required speed and manoeuvrability, had proved a difficult balancing act for RNLI technicians and the development of the Severn had taken longer than was originally planned. As with any new class of lifeboat, crews around the country were looking for proof that what the RNLI had to offer as a replacement to the trusty Arun did, in fact, represent real progress and was worth the extra effort and expense of caring for a more highly-strung machine. The story of the performance of RNLB *Michael and Jane Vernon* off the outlying Shetland island of Bressay on the afternoon of Wednesday 19 November did much to provide the proof the crews were looking for.

As he watched the dimly-lit white hull of the 3,600-ton refrigerated cargo ship *Green Lily* glide past him out of the harbour entrance, the harbour master must have wondered if her skipper's determination to resume his voyage in such weather would come back to haunt him. As soon as she turned south into the still relatively sheltered Bressay Sound, her bow began to ride up the 15ft swells and lurch heavily into the trough beyond. The forecast had left no one in any

The factory ship *Pionersk* in her final resting place south of Lerwick, Shetland, after Hewitt Clark had taken off 64 of her crew in storm-force winds in October 1994. He was awarded the RNLI Silver Medal. (*Graeme G. Storey*)

doubt that Shetland was about to experience a storm that even its gale-hardened locals might consider extreme.

By 8am on 19 November 1997, the wind, a south-easterly Force 10 to 11, was showing that it could and would live up to its billing. It came screaming off the North Sea, sending salt spray high over the rocky headlands, pummelling the grass-covered inland wastes with outraged fury and goading breakers the size of town halls on towards their destruction against the granite cliffs of Shetland.

To be at sea that day in any vessel would be an ordeal and there were certainly some who wondered what the Croatian captain and 14-man crew of the *Green Lily* were going through just then. They found out all too soon. At 8.45am her skipper informed the Shetland Coastguard that he had developed engine problems and was about 14 miles south-east of Lerwick, drifting back towards Bressay at about 2 knots.

There was still time to save the ship with tugs if her engines could not be restarted. Two were tasked for immediate response, the 38m *Tystie* from the Sullom Voe terminal and the 64m *Gargano* from Lerwick. A third, the *Maersk Champion*, would set out later from Lerwick once she had discharged a cargo. Hewitt Clark was relieved to hear that neither the lifeboat nor the Coastguard rescue helicopter was required, but he did not envy the tug skippers their task.

The *Gargano* was first to arrive on the scene at 11.15am and immediately set about passing a towline to the cargo ship which was wallowing in the mountainous swell, still without power. A little more than half an hour later the tug was able to radio that she had the ship in tow and was heading for Dales Voe base, north of Lerwick, where she would rendezvous with the *Tystie*. Then, 50 minutes later, the *Gargano's* master came on the radio again to say that the towline had parted and that it would take him at least another hour to get the *Green Lily* once more under tow. This time lives were seriously at risk; the freighter was considerably closer to a lee shore now and there was no alternative but to scramble *Rescue Lima Charlie*, the Coastguard Sea King helicopter, and to ask Lerwick lifeboat to put out.

Both helicopter pilot and lifeboat coxswain knew, even if their job was just to stand by, that they and their craft would be unlikely ever to face a more challenging battle with the elements. Every member of the six-man lifeboat crew was strapped into their seat inside the wheelhouse as Hewitt Clark took the lifeboat at no more than 10 knots out of Bressay Sound to take on the full force of the gale.

They were meeting the Force 11 winds and 40ft waves head-on as they proceeded south-east to clear Bard Head, the southern tip of Bressay. Time was of the essence as the casualty, still not under tow, was reported to be only 1½ miles from the shore. It was not until he turned eastwards, with the seas now on his beam, that Hewitt Clark could open the throttles. He was delighted to see how well his new Severn class coped with the weather, even at 20 knots. In the distance he could see the rescue helicopter stationary in the sky and he knew she must be over the casualty. He could also hear the Coastguard urgently asking the master of the *Green Lily* to be ready to release the anchors and to prepare as many of his crew as possible for evacuation.

By now it was about 1.40pm and the second tug, *Tystie*, had arrived. She wasted no time in working her way downwind towards the freighter's bow and successfully passing a heaving line to the two men working on her foredeck. But more precious minutes were then wasted as the men struggled to haul the heavy towing line aboard by hand, instead of using the windlass. They were still trying to get the line aboard when the lifeboat arrived on scene. The coxswain and crew could see the two figures waltzing crazily about the deck as their ship, beam on to the 40ft seas, rolled violently, seemingly determined to shake off her tether.

A last-ditch effort by the tug *Tystie* to tow the *Green Lily* away from danger. Soon the lifeboat, right, would have to get involved. (*RNLI*)

At last a third man was sent forward to assist with the desperate tug-of-war and eventually together they heaved the tail of the towline aboard and made it fast. The line of white spray marking the shore was now alarmingly close and the seas which broke over the *Green Lily* were stacking up higher than ever as they met waves reflected off the shore. The *Tystie* had not even begun to put weight on the towline before a mammoth sea lifted the freighter's bow high to port while the tug pitched in the opposite direction. The strain on the line was too great and it snapped.

This was the moment when Hewitt Clark and Captain Norman Leask, the helicopter pilot, knew that they were no longer spectators but lead players in the drama unfolding. The ship was less than a mile from the rocks and it was imperative she let go her anchors immediately. But where was the urgency on board? In spite of strong advice coming from the lifeboat, it was another 15 minutes before the starboard anchor eventually rattled into the water.

The single anchor brought the ship's head round into the wind and reduced the drift but did not halt it entirely. It was quite obvious by her violent motion and the seas breaking over her that a helicopter rescue was out of the question. Hewitt Clark began to weigh up whether his chances were any better. He knew that he would have to give it a go. Any approach would have to be on the lee side of the ship, even if that meant manoeuvring in the ever-decreasing space between the casualty and the shore. He could not help remembering how the coxswain of the Penlee lifeboat had been faced with a similar situation back in 1981 when everyone lost their lives; but at least this was in daylight and he had a much more nimble boat.

Clark and his second coxswain, Richard Simpson, moved from the shelter of the wheelhouse up to the steering position on the flying bridge above it. Meanwhile, three men, Brian Laurenson, Ian Leask and Michael Grant, made their faltering way forward to the starboard shoulder of the lifeboat while Peter Thomson took up a position amidships on the same side. The *Green Lily*'s skipper was urged to get his crew ready for evacuation.

The lifeboat moved gradually in to within 30ft of the ship's port side. Using his engines and his helm with utmost concentration, Hewitt Clark held the lifeboat in that position as the very confused seas lifted her high above the freighter's deck, then plunged her 50ft into a trough, deep below the waterline. It could not have been more obvious that he was now ready to come in to

Lerwick lifeboat alongside the *Green Lily*. The cliffs of Bressay are only 200 yards away. (*RNLI*)

take men off, but nobody appeared on deck and all the while the shore was looming larger.

Eventually, half a dozen men appeared in lifejackets and carrying suitcases. The lifeboat moved in without further delay. Not every approach Hewitt Clark made was going to be successful; sometimes he had to pull away because the superstructure of the two vessels threatened to collide as they rolled towards each other, at other times, just as he got close, the lifeboat would surge heavenwards, putting her crew at the level of the ship's rigging.

Sometimes a survivor was not ready to come forward as the lifeboat made her run in and sometimes the rescuers found a man and his luggage too great a weight to haul aboard before the coxswain needed to pull clear. The luggage itself was occasionally thrown on board without its owner, hampering the rescue yet further.

At one moment the lifeboat became trapped alongside and, fearing that he would be crushed at any moment, Clark went full ahead port, full astern starboard to wrench himself clear. It tore a stanchion, the forward toe rail and a piece of the fendering away, but it probably saved the lifeboat. Michael Grant had been attached by his lifeline to the guardrail, which was now hanging over the side, but he and Ian Leask managed to grab the fallen stanchion, haul it inboard, unclip the lifeline and resecure it further aft.

Among these abandoned attempts, there were occasions when both decks came level long enough for a man to be unceremoniously dragged aboard the lifeboat. There were five safe inside the wheelhouse when Hewitt Clark noticed that the ship's bow was beginning to turn into the wind and he was losing the lee on her port side. He and his crew had not been aware of a daring manoeuvre which had been performed by the *Maersk Champion*, the third tug to reach the scene. In a remarkable display of seamanship, her master had driven close to the *Green Lily*'s bow and grappled her anchor cable. He then attached it to his towline and was holding the freighter away from the rocks, now only 200 yards away.

With the ship's head to wind, it would have been suicidal to keep trying to get the lifeboat alongside, so the coxswain steered clear. The helicopter, however, now had a better chance of getting its winchman, Billy Deacon, down as the ship was pitching rather than rolling in her new position. The pilot and his crew worked fast. Ten men still had to come off the ship. Billy Deacon was lowered to the deck where, two at a time, he loaded them into the strop and prepared the next ones for the lift.

During this operation, which took little more than ten minutes, the ship's anchor cable parted, her bow swung round to starboard, and she began to drift rapidly towards the cliffs. By now all ten seamen were aboard the helicopter and it only remained to lower the wire one last time to collect the winchman. The

The helicopter scours the sea for a sighting of its lost winchman, Billy Deacon, who had helped ten men to be lifted off the *Green Lily* before he himself was swept into the sea. He was awarded the George Medal posthumously.

ship was back broadside to the sea and, to their horror, the helicopter crew saw a massive wave break over the *Green Lily* just as the wire descended. Billy Deacon was swept into the sea and then the ship appeared to rise and fall heavily on her port side, seemingly on top of the winchman. Almost immediately afterwards her stern hit the rocks.

As the helicopter hovered above the scene, her crew scouring the water for a sign of their colleague's orange overalls and yellow helmet, the winch wire snagged on the rigging of the wrecked ship. Only the winch operator's swift

The Gold Medal crew from Lerwick, left to right: Ian Leask, Michael Grant, Brian Lawrenson, Peter Thorogood, Richie Simpson and Coxswain Hewitt Clark. (*Dennis Coutts*)

action to cut the wire saved the helicopter, her crew and the survivors from plummeting to their deaths.

What on earth should the lifeboat coxswain do now? No vessel could possibly survive among the rocks where the lost winchman would be. The Coastguard had asked him to stay there and search but both he and the helicopter pilot were sure that they could only be looking for a body. Hewitt Clark took the difficult decision to land his survivors rather than to risk them in those appalling conditions any longer. Thanks to the speed of the lifeboat, the survivors were put ashore at 3.20pm, only some 40 minutes after the last one had come off the ship.

Meanwhile, the search continued and without hesitation the coxswain and his crew headed back to the scene of the shipwreck. The weather was no better and the light was beginning to fail. An RAF helicopter from Lossiemouth had joined in the search by the time the lifeboat was back. Now Hewitt Clark went as close in as he dared to the shore. Already, the *Green Lily* was disintegrating, shedding cargo, steel hatch covers, pallets, ropes and oil into the water around her and making it all the more dangerous for the lifeboat. While she was searching close to the cliff, one huge wave broke over the lifeboat from astern, engulfing her and almost pitch-poling her upside-down. Hewitt Clark took that as a signal that he had taken enough risks with his crew's lives for one day and pulled away from the shore.

Billy Deacon's body was found the next day having been washed some 7 miles up the coast of Bressay. His widow later received the George Medal posthumously for his bravery and the RNLI also presented its Thanks of the Institution on Vellum to him and the rest of the helicopter crew. As well as the Gold Medal awarded to Hewitt Clark, the RNLI Bronze Medal went to each of his crew.

16

TORBAY, 13 JANUARY 2008

During a winter of many storms and much work for the RNLI, a cargo of timber shifts aboard a freighter off the south coast of Devon in Force 9 winds. After a helicopter lifts 12 of the 20-man crew from the severely listing ship, Coxswain Mark Criddle and his crew succeed in taking the remaining men aboard Torbay lifeboat in extremely dangerous circumstances. The Silver Medal is awarded to Mark Criddle for his courage, skill and determination.

The winter months of late 2007 and early 2008 will be remembered by many lifeboat crews throughout the UK and Ireland for a succession of low pressure weather systems which swept across all areas of the British Isles bringing inland floods and furious sea conditions. The relentless gales and rain did not seem to deter the growing number who seek pleasure and thrills around the coast, even in the depths of winter, and this is reflected in some of the emergencies lifeboats were called to in the period.

Scarborough lifeboat launched into a Force 6 with 15ft swell on 21 November 2007 to try to reach an exhausted surfer. The 14-tonne lifeboat became airborne on several occasions during the search, which ended when the surfer was able to struggle into the shallows on his own, although he needed hospital treatment once ashore.

Lifeboats from Tynemouth, Sunderland and Hartlepool, together with an RAF helicopter and shore-based Coastguard teams from Seaham and Sunderland were scrambled two days later after a radio distress call reporting that a pleasure craft had sunk and its three occupants had taken to a life-raft. After two hours of searching, the Coastguard ascertained that the radio signal had come from a transmitter on land and that they were the victims of a hoax.

There had been nothing imaginary about three people aboard a cabin cruiser that had been seen setting out from Whitby Harbour in atrocious conditions soon after midday on that same day. Volunteers at the lifeboat station, who were among the witnesses, tried to contact the ill-advised sailors by radio to warn them of the danger they were facing. Then as the boat tried to turn round, she

Whitby's Trent class lifeboat was called out to a cabin cruiser that had set out from Whitby Harbour in atrocious conditions in November 2007. (*Robert Townsend*)

capsized, throwing her crew into the sea. Minutes later the all-weather lifeboat was launched and she soon located two of the occupants about 330ft from the harbour's west pier. The third person was plucked from the sea by an RAF helicopter winchman. Tragically, and in spite of the rescuers' prompt response, none of the casualties survived.

On 1 December, unbeknown to the crews of Weston-super-Mare's Atlantic 75 and D class lifeboats, a kite-surfer, whose abandoned kite was spotted in the waters of the Severn Estuary, had got ashore safely. Until the news of his safety reached them, the lifeboats had persevered with their search in a Force 5, very choppy conditions and a blinding hailstorm.

Arbroath lifeboat station on Scotland's east coast recorded the first coastal rescue of 2008 when both their all-weather and inshore lifeboats responded to a call on New Year's Day from some anglers, one of whom had slipped on rocks and broken an ankle. In spite of thick fog, the crews located their casualty, went ashore to administer first aid and then ferried the man on a stretcher aboard the inflatable D class to the larger lifeboat, which carried him to an ambulance waiting in the harbour.

In the same month, a pair of shoes left on a beach in Sunderland and a car abandoned by the sea at New Quay in Wales both led to hazardous but fruitless night-time searches by the towns' lifeboat crews. However, when a father arrived breathless and soaking wet at the Staithes and Runswick lifeboathouse on 8 January, having left his family stranded by a rising tide on the Yorkshire coast to get help, the lifeboat crew, who had just completed their Sunday morning exercise, were able to speed to the rescue and return the man's wife, two children and their dog to safety within half an hour.

With so many calls on lifeboats to people at leisure, there is a misconception held by some observers that the RNLI is losing its relevance to the world of commercial shipping. How often in this high-tech age, they ask, are lifeboats of any use to cargo vessels, ferries or even fishing boats? In the same three months that RNLI crews were so busy bringing pleasure seekers ashore, their numerous services to those who earn a living from the sea ranged from an incident when the lifeboat itself became a casualty to the most daring mid-channel rescue of recent decades.

On 29 January 2008, the skipper of a rigid inflatable workboat (RIB) and his two passengers found themselves unable to manoeuvre away from rocks close to the entrance to the harbour of Rathlin Island, off the north coast of Ireland. In heavy sea conditions the Portrush Severn class put out to their assistance, but while the coxswain attempted to get close enough to pass a towline, the lifeboat was picked up in the swell and dumped on to the rocks. While the three occupants of the RIB were able to get ashore across the rocks, the lifeboat stuck fast. Her crew were taken off but the £1.4 million lifeboat remained stranded for several weeks and was subsequently deemed to be beyond economic repair.

Such a vastly expensive mishap is a sobering reminder of the risks lifeboat coxswains must sometimes take close in to shore for the sake of stranded seafarers. Only two months earlier on 22 November 2007, Howth's Trent class and D class inflatable lifeboats, stationed just north of Dublin, were called out at 4am to the aid of a fishing vessel that had hit a rock in Balscadden Bay and was sinking. When they arrived they found the four-man crew clinging to the rock and their vessel disappearing fast beneath the waves. The lifeboat crews successfully retrieved the men from the rock and ferried them back to dry land.

On the early morning of 30 January 2008, another fishing crew was rescued close to rocks, this time under an Orkney cliff face. They were forced to abandon their 10.7m vessel when she struck bottom and began to take on water. It was still dark when the Kirkwall lifeboat reached the scene, but the illuminating flare that they fired lit up the reflective strips on the life-raft which the fishermen had taken to. They were pinned under the cliffs, but the lifeboat coxswain edged as close as he dared until a heaving line could be thrown to the men. They caught it, made it fast to their life-raft and were pulled clear and to safety aboard the lifeboat.

When the 12m fishing vessel *Matthew Harvey* was returning with her catch to Scarborough on 3 January 2008, her skipper wisely asked whether an escort was available. He had not been working out of the Yorkshire port for long and in a Force 8 and a huge, 26ft swell he needed some local guidance. Scarborough's all-weather Mersey class was soon heading out to meet him through the hefty, pounding surf and stood by as the fishing boat made her dash for the entrance and safety.

Arranmore's 25-knot Severn class lifeboat, stationed on the coast of County Donegal, had 60 miles to cover to reach a trawler with 16 Spanish crew on board that had caught fire on 19 January. Non-essential crew had been airlifted off by an Irish Coast Guard helicopter by the time the lifeboat arrived and six of the remaining seven took refuge on board the lifeboat while they waited for the arrival of an Irish naval vessel which would help the skipper put out the fire. With the fire extinguished the crew went back on board and with power restored the lifeboat left the scene. The trawler then called her back as one of the crew needed medical attention. He was taken by the lifeboat for hospital treatment on the Irish mainland and by the time the crew were back at their station, they had spent more than 12 hours at sea.

A storm which engulfed northern Britain on 31 January and 1 February led to lifeboats launching into some of the most extreme conditions they are ever likely to face. Two lifeboats, one from Fleetwood, the other from Lytham St Anne's, put out in Force 9–10 winds when a ferry, *Riverdance*, carrying lorries and trailers between Warrenpoint in Northern Ireland and Heysham on the Lancashire coast, developed a frightening 50-degree list after a huge wave had shifted her cargo. Two rescue helicopters winched off 14 passengers and crew as the lifeboats stood by in the driving sleet and mountainous seas. Two hours later the ferry was aground off Blackpool and the remaining crew, who first thought they would be safe, issued a Mayday call as their ship began to heel over even further. Fleetwood lifeboat set out again at 5am and stood by as the men were winched off by helicopter.

Meanwhile, during the same storm the trawler *Spinningdale* had been flung against a cliff on the remote island of St Kilda, some 40 miles west of the Hebrides. She had 14 men on board and was taking on water. The Severn class lifeboat from Stornoway set out in violent Storm Force 11 winds and despite the appalling conditions was able to reach the island in time to be on standby as the Coastguard helicopter hoisted the Spanish crew to safety.

A few weeks earlier, on the evening of Sunday 13 January 2008, in what might for the crews of Torbay and Salcombe lifeboats have turned out to be another task of standing by, albeit in ferocious seas, a highly dangerous operation became necessary, both for the rescue helicopter and for the lifeboats.

When Captain Arvanitis Charalampo steered the 6,500-tonne, Greek-registered cargo ship MV *Ice Prince* out of the Swedish port of Iggesund and headed south through the Gulf of Bothnia towards the southern Baltic, he knew he would be lucky to avoid some uncomfortable weather before he was out of European waters. Two-fifths of his cargo of timber, destined for Alexandria in

Stornoway's Severn class lifeboat puts out in a violent Storm Force 11 to the aid of a trawler 40 miles away, pinned against the cliffs of the remote island of St Kilda, February 2008. (*RNLI*)

Egypt, was in full view on the deck in front of him; bundles of 10m planks, tightly stowed and covered with plastic sheeting to guard against their notorious habit of shifting in rough seas, especially once they had become wet.

By the time he reached the English Channel, the weather was more than uncomfortable. A southerly gale, Force 8–9, had built steadily throughout the

The Seatruck ferry *Riverdance*, stranded 5 miles off Blackpool after her crew and passengers had been airlifted to safety, 2008. (*RNLI*)

day and his ship was rolling with manic energy as huge seas pushed their way across her westward path. Before long, he and his 19 crew became sickeningly aware that their ship was losing her equilibrium; as she heeled first to starboard, then to port, she was not returning to a fully upright position. She tended to stay over to port, at first only by a few degrees, but then the list increased to some

25 degrees. There was no doubt from anyone aboard that her cargo was to blame. If that was not bad enough in such a severe gale, when the ship suddenly lost all power, the captain did not hesitate to send out a Mayday. He was drifting helplessly some 34 miles off the Devon coast and for all he knew, was about to turn turtle.

The response by rescuers was immediate. The Coastguard helicopter *India Juliet*, based at Portland, was scrambled and both Salcombe and Torbay lifeboats put out into the pitch dark and the gale. The time was 7.45 in the evening. Mark Criddle, coxswain of the Torbay Severn class *Alec and Christina Dykes*,

Torbay's Severn class lifeboat, *Alec and Christina Dykes*. (RNLI)

encountering 13ft breaking waves, which had been whipped up by a fierce battle of wind against tide, throttled back to 16 knots and told his crew to strap themselves into their seats and to expect a very bumpy ride for the two hours it would take to reach the cargo ship.

Coxswain Criddle tried but failed to make contact with the *Ice Prince* by VHF radio. Brixham Coastguard, which was in contact with the ship, relayed the news that her list had increased to 45 degrees and her only source of power was emergency batteries. One of the crew had a broken leg. To Criddle the situation sounded desperate. The ship could capsize at any minute. He needed to get there faster – so he opened the throttles. At 20 knots the lifeboat was leaping, airborne, from one wave to the next as the crew braced themselves for one crash landing after another. Nothing whatsoever was visible through the wheelhouse windows in the spray and darkness – the coxswain was relying entirely on his instruments, especially the radar, for a safe passage.

The helicopter was first to arrive at the scene at 9pm. The pilot, Captain Kevin Balls, planned to winch the non-essential crew off the ship. Knowing from the Coastguard that Torbay lifeboat was only four minutes away, he set to work with the extremely difficult task of lowering his winchman down to the port wing of the *Ice Prince*'s bridge. The entire ship's company had mustered on the opposite starboard wing but the pilot needed to stay over the lower port side to be able to see the superstructure and aerials as they swung menacingly close to the rotor blades with every roll of the ship.

They got the winchman down and a group of nervous lifejacket-clad figures slid their way down the steeply sloping bridge to the opposite side of the ship. The winching took three-quarters of an hour, in which time 12 men were taken aboard the helicopter. On two occasions during the operation the helicopter nearly collided with the ship and was only saved by the pilot applying full power to pull clear in the nick of time. The hi-line, used for guiding the winch strop down on to the bridge, broke on three occasions and a new one had to be manoeuvred into the winchman's hands after each breakage.

When Torbay lifeboat arrived she had immediately taken up a position with her head to weather off the casualty's port quarter, her coxswain holding the boat in position from the upper steering position above the wheelhouse and crewmember Darryll Farley beside him, straining to keep the searchlight trained on the ship to give as much light as possible to the helicopter crew. All this was

The *Ice Prince* wallows in heavy seas
after all her crew have been evacuated.
She eventually sank, stern-first, 26 miles
south-south-east of Portland Bill. (*RNLI*)

done without any communication between lifeboat and helicopter. For some reason, the lifeboat mechanic Matthew Tyler had not been able to raise them on the VHF.

Salcombe's Tyne class *Baltic Exchange II*, under the command of Coxswain Marco Brimacombe, reached the scene at 9.30pm. She, too, took up a position close to the Torbay lifeboat and provided additional illumination. She was also

unable to make radio contact with the helicopter. Another vessel, the Royal Navy's frigate HMS *Cumberland*, which had been at anchor in Torbay, had also set out in response to the Mayday and when she arrived towards the end of the winching operation she placed herself upwind of the *Ice Prince* in order to provide a lee.

When, at about 10pm, the helicopter rose clear of the cargo ship, the winchman safely back aboard, there was some puzzlement among both lifeboat crews. The Torbay coxswain had earlier asked the Coastguard whether a second helicopter could be deployed but no answer had come back. In fact, the Aeronautical Rescue Co-ordination Centre at RAF Kinloss had refused the request, but *India Juliet's* pilot, having heard the request being passed on by the Coastguard, assumed another aircraft was on its way and turned for home with 12 rescued crewmen on board.

The situation was now becoming increasingly urgent for the people still aboard the ship. Criddle could make them out quite clearly huddled together on the bridge wing. Fortunately, he was able to make radio contact with Captain Charalampo, although the captain's rudimentary English made the exchanges difficult. In fact, the captain had now ordered all remaining men, eight including himself, to abandon ship. The angle of list made it impossible to use either of the ship's lifeboats and he was asking whether the coxswain was prepared to help them. Criddle agreed without hesitation and asked the captain to ensure his men were wearing immersion suits and lifejackets with lights and to tell them to assemble at the stern of the ship where he would make his approach. Meanwhile, the lifeboat crew donned helmets and safety harnesses, rigged fenders on the port shoulder and prepared rescue strops, including one on the end of a line, in case anyone fell in the water. Then they made their way to the foredeck, all except Darryll Farley who was needed at the searchlight.

Any approach to the casualty's stern was fraught with danger. The ship was drifting beam-on to the wind at 3 knots. There would be no shelter from the elements as the lifeboat came in close and with the necessary pinpoint accuracy: too far to port and the submerged gunwale would thrust its way up under the lifeboat's hull; too much the other way and the starboard quarter would come down like a guillotine, crushing the crew on the foredeck. To make it even more difficult for the coxswain, a large anchor was housed on the stern, giving very little free space to come alongside.

Using hand signals, the coxswain indicated that he wanted the ship's crew to come, one at a time, from their position on the high side of the stern, down the steeply sloping deck to the port side where the sea was washing over the deck. That took courage in itself, but the first man eventually slithered to the required position and the lifeboat made her approach. Everything went perfectly. The lifeboat's port shoulder came close enough to the *Ice Prince*'s stern for the man to be able to step across into the arms of the lifeboat's two deputy second coxswains, Roger Good and John Ashford, who were both forward of the pulpit in the bow. He was bundled back to the wheelhouse by the other three crewmembers as the lifeboat pulled clear to prepare for her next approach.

Although it took several attempts to get close enough again in the turbulent water around the stern without actually making contact, Criddle, using all his concentration and skill, managed it twice more to allow two more survivors to come aboard. Coming in close for the fourth man, the lifeboat rolled suddenly and the two vessels crashed together. Every crewman on the lifeboat foredeck was thrown off his feet and the lifeboat's bow fendering hung in shreds.

The man himself lost his hold and slid down the deck and into the water. The coxswain reversed clear, terrified that the man was about to be crushed between the two vessels. As the crew on the bow picked themselves up, they were hugely relieved to see a surge of water push the man back to his original perch. On the next approach the lifeboat crew grabbed him and he was led, unhurt, to safety.

There was a new problem now; the remaining four men had taken fright and would not come down to the position where they could be got aboard the lifeboat. In spite of numerous approaches and screamed persuasion and gesticulation from the lifeboat crew, the men would not budge. In desperation, the coxswain on one occasion used power to push his bow against the stern to hold his position to make it easier for the men but all it achieved was further damage to the lifeboat's bow.

All the lifeboat could do was to keep trying. Time and again she drew close, then backed away until, at last, the men began to move. Again the coxswain edged forward and this time a man half jumped, half fell towards the lifeboat. Only by the speed and strength of the five men on the foredeck was he saved from falling between the two boats. They grabbed him and hauled him up over the side, then hurried him aft to join his other shipmates.

A grateful Captain Arvanitis Charalampo, right, with Torbay Coxswain Mark Criddle after the captain and his crew had been rescued from the MV *Ice Prince*. The ship, carrying a cargo of timber, began listing in heavy seas off the Devon coast in January 2008. (*Nigel Millard*)

The last three men were almost as difficult to get off. Each needed several more approaches and each had to be manhandled aboard, the foredeck crew risking being crushed every time they stepped forward to grab a survivor. At last they could pull away for the final time and prepare for the passage home. It had taken them an exhausting one-and-three-quarter hours with more than 50 manoeuvres alongside to get all the men off. Their almost impossible task had been helped considerably by Salcombe lifeboat which had kept her position throughout, giving extra illumination from her searchlight. HMS *Cumberland* had also been a reassuring presence as well as providing a lee.

A check of the lifeboat's damaged bow revealed that the hull was still intact. Of the survivors, one looked groggy and crewmember Alex Rowe, who is a doctor, suspected a heart condition. Another showed signs of broken ribs. In spite of their buffeting, the lifeboat crew were unhurt except for Darryll Farley who had cut his head on the searchlight in his battle to keep it on target.

Torbay lifeboat reached her home port at 1.15am, having been at sea for five-and-a-half hours. Captain Balls, the helicopter pilot, only discovered when he arrived back at Portland to land his 12 survivors that another helicopter had not been sent to the scene. He immediately refuelled and returned to the ship and arrived just as the last survivor had made it aboard the lifeboat.

No one could have known it at the time, but the *Ice Prince* remained in her perilous listing state for two more days. The weather prevented any salvage attempt and she eventually sank, stern-first, 26 miles south-south-east of Portland Bill. Much of the timber that had been on her deck eventually fetched up in spectacular quantity on the beaches of Sussex.

Mark Criddle did not get back home until after four in the morning, following his extraordinary night at sea. When he left the house at seven the previous evening, he had been doing the children's laundry. He told reporters later that the first thing his wife Melanie asked him as he clambered wearily into bed was: 'Did you bring that washing in?'

Once the full story of that night was revealed, the RNLI saw fit to award Mark Criddle the Silver Medal for his courage, leadership, seamanship, initiative and determination. His crew – Roger Good, John Ashford, Matthew Tyler, Nigel Coulton, Alex Rowe and Darryll Farley – all received the official Thanks of the Institution on Vellum for the considerable part they played in the safe return of the crew of the *Ice Prince*.

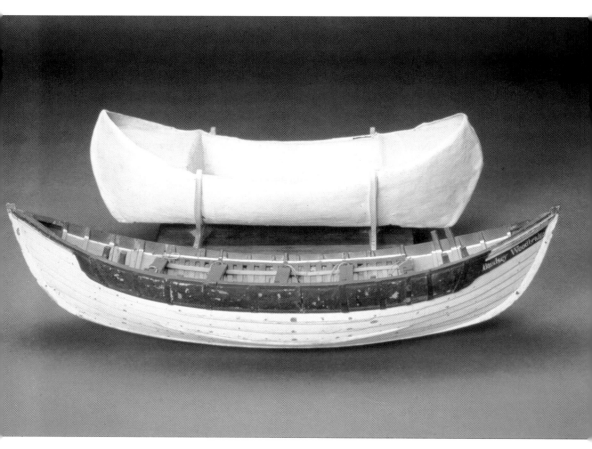

The origins of the lifeboat service: the model in the background of a self-righting design was built in tin by William Wouldhave in 1789 as his winning entry in a competition to find the best boat for saving lives. The model in the foreground is of the 'Original', a design developed by Henry Greathead, using a combination of Wouldhave's and other competition entrants' ideas and from which the earliest lifeboats were built.

BIBLIOGRAPHY

PRIMARY SOURCES

British Library, Newspaper Library – various contemporary local newspapers
The Grahame Farr Archives
Royal National Lifeboat Institution Archives

SECONDARY SOURCES

Bourke, Stephanie, *The Hamilton Family & The Making of Balbriggan* (Balbriggan and District Historical Society, 2004)

Cameron, Ian, *Riders of the Storm* (London, Weidenfeld & Nicolson, 2002)

Cox, Barry, *Lifeboat Gallantry* (London, Spink and Son, 1998)

De Courcy Ireland, J., *Wreck and Rescue on the East Coast of Ireland* (The Glendale Press, 1983)

Denton, Tony, *Lifeboat Enthusiasts' Society Handbook 2008*

Farr, Grahame, *Wreck and Rescue in the Bristol Channel 1* (D. Bradford Barton, 1966)

Howarth, Patrick, *Lifeboat: In Danger's Hour* (London, Hamlyn, 1981)

Kelly, Robert, *For Those in Peril* (Shearwater Press, 1979)

Kipling, Ray and Susannah, *Strong To Save* (Patrick Stephens Ltd, 1995)

Miller, Allen, *The Great Lifeboat Disaster of 1886* (Sefton Council, Leisure Services Dept (Libraries), 2001)

Morris, Jeff, *The Story of Fraserburgh Lifeboats*, 2003

Ortzen, Len, *Famous Lifeboat Rescues* (Arthur Baker Ltd, 1971)

Royal National Lifeboat Institution, *The Lifeboat Journal* and *Annual Reports*

Sagar-Fenton, Michael, *Penlee, The Loss of a Lifeboat* (Bossiney Books, 1991)

Skidmore, Ian, *Lifeboat VC* (David & Charles, 1979)

Vince, Charles, *Storm on the Waters* (Hodder & Stoughton, 1946)

Warner, Oliver, *The Lifeboat Service* (London, Cassell, 1974)

INDEX